AF362030

New Insights in (Poly)Phenolic Compounds

New Insights in (Poly)Phenolic Compounds

From Dietary Sources to Health Evidences

Editors

Cristina García-Viguera
Raúl Domínguez-Perles
Nieves Baenas

MDPI • Basel • Beijing • Wuhan • Barcelona • Belgrade • Manchester • Tokyo • Cluj • Tianjin

Editors
Cristina García-Viguera
National Council for Scientific
Research (CEBAS-CSIC)
Spain

Raúl Domínguez-Perles
National Council for Scientific
Research (CEBAS-CSIC)
Spain

Nieves Baenas
University of Murcia
Spain

Editorial Office
MDPI
St. Alban-Anlage 66
4052 Basel, Switzerland

This is a reprint of articles from the Special Issue published online in the open access journal *Foods* (ISSN 2304-8158) (available at: https://www.mdpi.com/journal/foods/special_issues/ Polyphenolic_Dietary_Sources_Health).

For citation purposes, cite each article independently as indicated on the article page online and as indicated below:

LastName, A.A.; LastName, B.B.; LastName, C.C. Article Title. *Journal Name* **Year**, *Article Number*, Page Range.

ISBN 978-3-03943-378-0 (Hbk)
ISBN 978-3-03943-379-7 (PDF)

Cover image courtesy of Cristina García-Viguera.

Contents

About the Editors . . . . **vii**

Raúl Domínguez-Perles, Nieves Baenas and Cristina García-Viguera
New Insights in (Poly)phenolic Compounds: From Dietary Sources to Health Evidence
Reprinted from: *Foods* **2020**, *9*, 543, doi:10.3390/foods9050543 . . . . **1**

Estrella Galicia-Campos, Beatriz Ramos-Solano, Mª. Belén Montero-Palmero,
F. Javier Gutierrez-Mañero and Ana García-Villaraco
Management of Plant Physiology with Beneficial Bacteria to Improve Leaf Bioactive Profiles
and Plant Adaptation under Saline Stress in *Olea europea* L.
Reprinted from: *Foods* **2020**, *9*, 57, doi:10.3390/foods9010057 . . . . **5**

María C. Sánchez, Honorato Ribeiro-Vidal, Begoña Bartolomé, Elena Figuero,
M. Victoria Moreno-Arribas, Mariano Sanz and David Herrera
New Evidences of Antibacterial Effects of Cranberry Against Periodontal Pathogens
Reprinted from: *Foods* **2020**, *9*, 246, doi:10.3390/foods9020246 . . . . **21**

Rocío González-Barrio, Vanesa Nuñez-Gomez, Elena Cienfuegos-Jovellanos,
Francisco Javier García-Alonso and Mª Jesús Periago-Castón
Improvement of the Flavanol Profile a nd t he A ntioxidant C apacity o f C hocolate U sing a
Phenolic Rich Cocoa Powder
Reprinted from: *Foods* **2020**, *9*, 189, doi:10.3390/foods9020189 . . . . **41**

Vicente Agulló, Raúl Domínguez-Perles, Diego A. Moreno, Pilar Zafrilla and
Cristina García-Viguera
Alternative Sweeteners Modify the Urinary Excretion of Flavanones Metabolites Ingested
through a New Maqui-Berry Beverage
Reprinted from: *Foods* **2020**, *9*, 41, doi:10.3390/foods9010041 . . . . **53**

Javier Marhuenda, Silvia Perez, Desirée Victoria-Montesinos, María Salud Abellán,
Nuria Caturla, Jonathan Jones and Javier López-Román
A Randomized, Double-Blind, Placebo Controlled Trial to Determine the Effectiveness a
Polyphenolic Extract (*Hibiscus sabdariffa* and *Lippia citriodora*) in the Reduction of Body Fat Mass
in Healthy Subjects
Reprinted from: *Foods* **2020**, *9*, 55, doi:10.3390/foods9010055 . . . . **63**

Javier Marhuenda, Silvia Perez-Piñero, Desirée Victoria-Montesinos,
María Salud Abellán-Ruiz, Nuria Caturla, Jonathan Jones and Javier López-Román
Correction: Marhuenda, J., et al. A Randomized, Double-Blind, Placebo Controlled Trial to
Determine the Effectiveness a Polyphenolic Extract (*Hibiscus sabdariffa* and *Lippia citriodora*) in
the Reduction of Body Fat Mass in Healthy Subjects. *Foods* 2020, *9*(1), 55
Reprinted from: *Foods* **2020**, *9*, 279, doi:10.3390/foods9030279 . . . . **75**

Francesca Oliviero, Yessica Zamudio-Cuevas, Elisa Belluzzi, Lisa Andretto, Anna Scanu,
Marta Favero, Roberta Ramonda, Giampietro Ravagnan, Alberto López-Reyes,
Paolo Spinella and Leonardo Punzi
Polydatin and Resveratrol Inhibit the Inflammatory Process Induced by Urate and
Pyrophosphate Crystals in THP-1 Cells
Reprinted from: *Foods* **2019**, *8*, 560, doi:10.3390/foods8110560 . . . . **79**

Juan Ángel Carrillo, M Pilar Zafrilla and Javier Marhuenda
Cognitive Function and Consumption of Fruit and Vegetable Polyphenols in a Young
Population: Is There a Relationship?
Reprinted from: *Foods* **2019**, *8*, 507, doi:10.3390/foods8100507 . . . . **93**

About the Editors

Cristina García-Viguera (PhD) is a graduate in Pharmacy (University Complutense of Madrid, 1985). Since 1988, she has been following a professional career at the Department of Food Science and Technology in CEBAS-CSIC (Spanish Research Council Institute), Murcia, Spain were she developed different HPLC techniques for food characterization, and obtained her PhD in Chemistry (University of Murcia, 1991). In 1992 she joined the Plant Science Department, at the University of Oxford, where she became familiar with GC-MS phytochemical analysis. Later on (1993) she worked at the Institute of Food Research (IRF), at Reading (UK). In 1994 she returned to CEBAS-CSIC where she developed a new and independent area of research in polyphenolic analysis in the field of food chemistry. In 1999 she became a Tenured Scientist (CSIC), followed by Scientific Researcher (2004), and Research Professor (2009). Now she leads a research group (Phytochemistry and Healthy Foods Lab) that has a prominent position with international recognition in the field of phytochemicals (phenolic compounds, glusosinolates, minerals and vitamin C) in food, and namely in: development of new beverages and foods (fresh or processed); chemical transformations of phytochemicals, resulting from the food industry processes or agricultural conditions; and biological properties (bioaccesibility, bioavailability and bioactivity) of these compounds. She presents in her CV over 150 original articles published in journals indexed in the Science Citation Index (h 55 scopus), five patents transferred to industries, several book chapters, numerous invited conferences and Congress communications, multiple I&DT projects, several supervisions of master's and PhD students; she is also co-founder of the CSIC spin-off Aquaporins & Ingredients SL (2009), and Aquaporins Dermoactivity (2018). In addition, she is CSIC Researcher-In-Charge for the Associated Unit "Calidad y Evaluación de Riesgos en Alimentación" (Univ. Politécnica de Cartagena- CSIC) 2016 (July)–2022 (July).

Raúl Domínguez-Perles (PhD) earned his bachelor's degree in Veterinary Medicine and DEA in Immunology (University of Murcia in 2001 and 2005, respectively). His research career is differentiated in two stages: first, research in Life Sciences and Health (2002-2008) and second, Doctoral and Postdoctoral research in the field of Food Science and Technology (2008-present), that constitute the body of his current research activity. It is noticed that the scope of the research undertaken during the second stage was favored by the skills acquired between 2002 and 2008 that allowed him to develop high impact investigations, applying an array of analytical tools, often not available in the Food Science and Technology lab. After obtaining his Ph.D. degree in 2011, he developed a postdoctoral stage of 24 months in the metabolomic lab of the Research Group at CEBAS, which enhanced his research capacities on bioavailability and bioactivity of bioactive compounds upon in vivo studies. In 2013, he started an international postdoctoral stage as Auxiliary Researcher at the University of Trás-os-Montes and Alto Douro (UTAD, Portugal) (40.5 months), where he developed autonomous research on the valorization of plant foods and their by-products as a source of phytochemicals, and developed the scientific coordination of the FP7-project EUROLEGUME (n° 613781), among other activities. In 2016, he was awarded with a 'Saavedra Fajardo' fellowship for the reincorporation of Doctors to the Spanish I+D+I system (CEBAS-CSIC) and in 2017 with a 'Juan de la Cierva of Incorporation'. These two contracts allowed him to establish his own research line as a senior researcher on the bioavailability and bioactivity of plant oxylipins. He has participated in 27 funding actions: 4 international, 9 national (1 as IP), and 4 regional (1 as IP) projects, and 2 networks of excellence, 6 company contracts, and 2 consulting activities (as a responsible researcher). In addition, he has published 84 SCI-articles, reaching an h-index of 24, as well as 8 book chapters with national (2) and international (6) editorials, 28 articles in scientific (non-SCI) journals, and 54 contributions in congresses and workshops, including 6 presentations as an invited speaker in plenary seasons of international congresses and 1 intervention in the Winter Conferences of the International PhD Program Agri-Chains (UTAD, Portugal). In respect to the supervision of research works, he has supervised 1 PhD project, 10 MSc theses, and 5 BSc theses.

Nieves Baenas (PhD) is a graduate in Agronomist Engineering. Subsequently, she obtained a master's degree in Nutrition, Technology and Food Safety in the year 2011 and since this year, she started her professional research career in the Department of Food Science and Technology in the "Centro de Edafología y Biología Aplicada del Segura" (CEBAS) of the Spanish Research Council (CSIC), Murcia, Spain, where she obtained her PhD. in Agricultural, Agro- Environmental and Food Resources and Technologies in the year 2016. Afterwards, she was awarded a Post-Doctoral Fellowship from the Martin Escudero Foundation (Spain) and completed this job in the Institute of Medicinal Nutrition from the University Medical Center Schleswig-Holstein, Campus Lübeck, Germany. She currently holds the prestigious Spanish fellowship Juan de la Cierva at the University of Murcia, within the Department of Food Science and Nutrition. Her works include studies on phytochemicals (phenolic compounds, glucosinolates, isothiocyanates and carotenoids) and dietary fiber for nutrition and health using HPLC-DAD-MS/MS, development of natural ingredients from industrial by-products for novel applications in the food and pharmaceutical industry, in vitro and in vivo studies of bioactivities for the preclinical development, studies of bioaccesibility and bioavailability of natural bioactive compounds, and detection of biomarkers and evaluation of gene expression in clinical trials for disease prevention.

 *foods*

Editorial

New Insights in (Poly)phenolic Compounds: From Dietary Sources to Health Evidence

Raúl Domínguez-Perles [1], Nieves Baenas [2] and Cristina García-Viguera [1,*]

[1] Phytochemistry and Healthy Foods Lab, Deptartment Food Science and Technology, CEBAS-CSIC, University Campus of Espinardo, 25, 30100 Espinardo, Murcia, Spain; rdperles@cebas.csic.es

[2] Department of Food Technology, Food Science and Nutrition, Faculty of Veterinary Sciences, University of Murcia, 30100 Espinardo, Murcia, Spain; nieves.baenas@um.es

*   Correspondence: cgviguera@cebas.csic.es; Tel.: +34-968-396200 (ext. 6304)

Received: 16 April 2020; Accepted: 28 April 2020; Published: 30 April 2020

Abstract: Nowadays, there is a gap between the theoretical bioactivity of (poly)phenols and their real influence in health, once ingested. Due to this, new studies, including in vitro and in vivo models that allow for exploring bioaccessibility, bioavailability, and bioactivity, need to be developed to understand the actual importance of consuming functional foods, rich in these plant secondary metabolites. Moreover, current new strategies need to be developed to enhance the content of these foods, as well as setting up new formulations rich in bioaccessible and bioavailable compounds. Altogether, it could give a new horizon in therapy, expanding the use of these natural functional compounds, ingredients, and foods in the clinical frame, reducing the use of synthetic drugs. As a result, the joint contribution of multidisciplinary experts from the food science, health, and nutrition areas, together with the industrial sector, would help to reach these objectives. Taking this into account, diverse studies have been included in this study, which comprises different strategies to approach these objectives from different, complementary, points of view, ranging from the enrichment of by-products in bioactive compounds, through different agricultural techniques, to the assimilation of these compounds by the human body, both in vitro and in vivo, as well as by clinical studies.

Keywords: (poly)phenols; bioactivity; bioavailability; inflammation; secondary metabolites; antibacterial; antioxidant; diet; fruit; vegetables

---

To date, the biological interest of bioactive phytochemicals in plant-based foods is widely accepted. Indeed, (poly)phenolic compounds are of particular interest. These benefits have been associated with their antioxidant, antimicrobial, anti-inflammatory, anticancer, and cardiovascular protection activities, which support the application of these compounds in pharmaceutical, food, and cosmetic industries. Nonetheless, there are still some gaps related to the bioaccessibility, bioavailability, and bioactivity of these molecules, once ingested by the diet, that require additional studies contributing to clarify these issues, including in vitro and in vivo studies. Further efforts on clarifying these aspects closely linked to the biological benefits associated with (poly)phenols will allow us to develop an understanding of the actual scope of consuming functional foods that are rich in these plant secondary metabolites. In this frame, the present study gathers seven articles, six research papers and one review that provide further insight into the effect of manufacturing and biological processes on the actual biological interest of (poly)phenols.

The first approach to the main objective of this study comes from the new challenge that arises from the global climate changes, which modulates soil salinity and increases the demand for water. In this hallmark, Garcia-Campos et al. [1] focused the research on the management of plant physiology with beneficial bacteria, to obtain a biological tool that contributes to improving leaf bioactive profile and plant adaptation under saline stress in olive trees. In this chapter, the authors report the improvement of the (poly)phenolic composition of plant by-products at the same time as increasing the fitness of the tree when using a beneficial bacteria strain. In this respect, the described results evidence the capacity of such a bacterial strain to trigger plant metabolism that targets several mechanisms simultaneously, concluding that several strains increase osmoprotectant activity and photosynthesis in olive tree leaves, affect the enzymatic and non- enzymatic antioxidant system, and increases the content of bioactive (poly)phenols in leaves. Rendering, in the end, a promising by-product, with enriched health bioactive compounds.

With the same objective, González-Barrios et al. [2] focused their research on the production system, to augment the content of the bioactive (poly)phenols, in this case, on a manufactured food, chocolate, by the use of a cocoa powder rich in these bioactive compounds. As a result, the authors obtained a formulation for new dark chocolate, enriched in flavan-3-ols and oligomeric procyanidins, with acceptable organoleptic characteristics, and high antioxidant properties, describing the main improvements made to manufacturing in the chapter.

Another aspect covered in this study is the activity of certain foods on several diseases, in which evidence was retrieved resorting to in vitro systems studies. A study by Sánchez et al. [3] describes the use of a rich (poly)phenolic cranberry extract against periodontal pathogens, as an alternative to classical antibiotics that could contribute to prevent antibiotic resistance. To achieve this objective, the authors studied the anti-bacterial activity, applying a validated in vitro biofilm model. The obtained results indicate that cranberry (poly)phenolic compounds interfere in the phase of bacterial adherence, preventing colonization by inhibiting the adherence of these pathogens to oral tissues. Besides, these extracts could reduce the incidence and severity of periodontal disease-related symptoms, by suppressing inflammatory cascades, caused by the immunological response to bacterial infection.

Following the same guideline, Oliviero et al. [4] studied the influence of polydatin and resveratrol, in the reduction of the inflammatory process induced by monosodium urate and calcium pyrophosphate in type-1 cells, demonstrating that both phenolic compounds are effective in inhibiting reactive oxygen species (ROS) and nitric oxide (NO) production, as well as crystal-induced inflammation. Due to this, the authors conclude that dietary supplementation with these compounds could support pharmacological treatments in patients affected by this inflammatory process and thus prevent the acute phase of the disease.

Likewise, studies of the efficacy or bioavailability of these compounds are also included in the present study, by the hand of diverse human clinical studies, concerning obesity. These studies are collected in two additional chapters. On one hand, Marhuenda et al. [5], have paid attention to the potential use of (poly)phenols to reduce obesity, studying the effectiveness of a blend formulated with *Hibiscus sabdariffa* and *Lippia citriodora* extracts for the treatment of obesity and/or overweight control in the absence of a controlled diet. The results demonstrate the association between the consumption of this blend and the reduction in body weight, Body Mass Index (BMI), and central fat mass, due to the possible changes in molecular pathways. On the other hand, Agulló et al. [6] have characterized a total of 16 flavanone urine metabolites, after the ingestion of a new maqui/citrus beverage, sweetened with sucrose (natural and high caloric), stevia (natural and low caloric) or sucralose (artificial and low caloric). The purpose of this study was to determine the influence of sweeteners in obesity, when added to a functional beverage, evaluating their possible interference with flavanones (phenolic bioactive compounds). The results indicate that both non-caloric additives provided a significantly higher urinary excretion for most compounds when compared to the traditional sweetener, sucrose. On this base, sucrose consumption could be reduced due to its evident influence on metabolic disorders, without providing any extra benefit in the absorption of bioactive (poly)phenols.

Finally, Carrillo et al. [7], reviewed the relationship between cognitive function and consumption of fruit and vegetable (poly)phenols in a young population, concluding that there is a positive effect of (poly)phenols related to neuroprotection and associated with the antioxidant and anti-inflammatory capacity of these compounds. This evidence emphasizes the importance of the intake of fruits and vegetables in maintaining normal cognitive function. Due to this fact, these bioactive compounds could also be considered nutraceutical supplements, useful as an alternative preventive therapy due to the lack of effective pharmaceutical treatments for age-related cognitive decline, implementing primary prevention actions in young people to minimize the cognitive impairment caused by age.

Author Contributions: All authors have made a substantial, direct and intellectual contribution to the work and approved it for publication. All authors have read and agreed to the published version of the manuscript.

Funding: This research received no external funding.

Conflicts of Interest: The authors declare no conflict of interest.

References

1.   Galicia-Campos, E.; Ramos-Solano, B.; Montero-Palmero, M.B.; Gutierrez-Mañero, F.J.; García-Villaraco, A. Management of Plant Physiology with Beneficial Bacteria to Improve Leaf Bioactive Profiles and Plant Adaptation under Saline Stress in Olea europea L. *Foods* **2020**, *9*, 57. [CrossRef] [PubMed]

2.   González-Barrio, R.; Nuñez-Gomez, V.; Cienfuegos-Jovellanos, E.; García-Alonso, F.J.; Periago-Castón, M.J. Improvement of the Flavanol Profile and the Antioxidant Capacity of Chocolate Using a Phenolic Rich Cocoa Powder. *Foods.* **2020**, *9*, 189. [CrossRef]

3.   Sánchez, M.C.; Ribeiro-Vidal, H.; Bartolomé, B.; Figuero, E.; Moreno-Arribas, M.V.; Sanz, M.; Herrera, D. New Evidences of Antibacterial Effects of Cranberry Against Periodontal Pathogens. *Foods* **2019**, *9*, 246.

4.   Oliviero, F.; Zamudio-Cuevas, Y.; Belluzzi, E.; Andretto, L.; Scanu, A.; Favero, M.; Ramonda, R.; Ravagnan, G.; López-Reyes, A.; Spinella, P.; et al. Polydatin and Resveratrol Inhibit the Inflammatory Process Induced by Urate and Pyrophosphate Crystals in THP-1 Cells. *Foods* **2019**, *8*, 560. [CrossRef] [PubMed]

5.   Marhuenda, J.; Perez, S.; Victoria-Montesinos, D.; Abellán, M.S.; Caturla, N.; Jones, J.; López-Román, J. A Randomized, Double-Blind, Placebo-Controlled Trial to Determine the Effectiveness a Polyphenolic Extract (Hibiscus sabdariffa and Lippia citriodora) in the Reduction of Body Fat Mass in Healthy Subjects. *Foods* **2020**, *9*, 55. [CrossRef] [PubMed]

6.   Agulló, V.; Domínguez-Perles, R.; Moreno, D.A.; Zafrilla, P.; García-Viguera, C. Alternative Sweeteners Modify the Urinary Excretion of Flavanones Metabolites Ingested through a New Maqui-Berry Beverage. *Foods* **2020**, *9*, 41.

7.   Carrillo, J.Á.; Zafrilla, M.P.; Marhuenda, J. Cognitive Function and Consumption of Fruit and Vegetable Polyphenols in a Young Population: Is There a Relationship? *Foods* **2019**, *8*, 507. [CrossRef] [PubMed]

Article

Management of Plant Physiology with Beneficial Bacteria to Improve Leaf Bioactive Profiles and Plant Adaptation under Saline Stress in *Olea europea* L.

Estrella Galicia-Campos, Beatriz Ramos-Solano, Mª. Belén Montero-Palmero, F. Javier Gutierrez-Mañero and Ana García-Villaraco *

Universidad San Pablo-CEU Universities, Facultad de Farmacia, Ctra Boadilla del Monte km 5.3, 28668 Boadilla del Monte, Madrid, Spain; e.galicia@usp.ceu.es (E.G.-C.); bramsol@ceu.es (B.R.-S.); mariabelen.monteropalmero@ceu.es (M.B.M.-P.); jgutierrez.fcex@ceu.es (F.J.G.-M.)
* Correspondence: anabec.fcex@ceu.es; Tel.: +34-91-3724785

Received: 8 November 2019; Accepted: 27 December 2019; Published: 7 January 2020

Abstract: Global climate change has increased warming with a concomitant decrease in water availability and increased soil salinity, factors that compromise agronomic production. On the other hand, new agronomic developments using irrigation systems demand increasing amounts of water to achieve an increase in yields. Therefore, new challenges appear to improve plant fitness and yield, while limiting water supply for specific crops, particularly, olive trees. Plants have developed several innate mechanisms to overcome water shortage and the use of beneficial microorganisms to ameliorate symptoms appears as a challenging alternative. Our aim is to improve plant fitness with beneficial bacterial strains capable of triggering plant metabolism that targets several mechanisms simultaneously. Our secondary aim is to improve the content of molecules with bioactive effects to valorize pruning residues. To analyze bacterial effects on olive plantlets that are grown in saline soil, photosynthesis, photosynthetic pigments, osmolytes (proline and soluble sugars), and reactive oxygen species (ROS)-scavenging enzymes (superoxide dismutase-SOD and ascorbate peroxidase-APX) and molecules (phenols, flavonols, and oleuropein) were determined. We found photosynthetic pigments, antioxidant molecules, net photosynthesis, and water use efficiency to be the most affected parameters. Most strains decreased pigments and increased osmolytes and phenols, and only one strain increased the antihypertensive molecule oleuropein. All strains increased net photosynthesis, but only three increased water use efficiency. In conclusion, among the ten strains, three improved water use efficiency and one increased values of pruning residues.

Keywords: olive; salinity; osmolytes; adaptation; secondary metabolism; plant growth promoting rhizobacteria (PGPR); net photosynthesis; oleuropein; water use efficiency (WUE)

---

1. Introduction

The traditional olive production system in the Mediterranean was developed in dry, farmed areas with trees spaced from 25 to 60 feet (7.6–18.3 m) apart, giving 12 to 70 trees/acre (30–173 trees/ha) [1]. Thus, olive trees are often under severe water deficit combined with high temperatures and high light intensities during the summer season. Moreover, the traditional olive production system in dry farms has many disadvantages such as low yields, delays before full production (15–40 years) and a very inefficient non-mechanical harvest [1]. In recent decades, the cultivation of the Spanish olive has undergone major technological changes associated with high-density or super high-density production systems, such as a reduction in the number of olive varieties, an increase in the density of the new plantations, an improvement of the harvesting machinery or orchard irrigation [2]. Therefore, this change from dry, farmed areas to irrigated cultivation has placed water stress or high salt concentration, as one of the main problems that olive cultivation is currently facing. In this agronomic framework, in which olive trees require irrigation, the possibility of producing in areas with high salt concentration or reduced water supply represents an important economic advantage.

Although olive resists a high degree of drought stress, the acclimation ability of olive plants to adjust to water deficit includes two mechanisms: avoidance and tolerance [3,4]. Among the innate acclimation mechanisms of plants are morphological and physiological leaf alterations; reduction of leaf size and number; biosynthesis and accumulation of compatible solutes (amino acids, proteins, sugars, methylated quaternary ammonium compounds, and organic acids); hormonal balance alteration (abscisic acid-ABA and ethylene); increase in ion efflux with high-affinity antiporters (salt overly sensitive-SOS1 and high-affinity potassium transporters-HKT); maintenance of reactive oxygen species (ROS) homeostasis, and decrease of photosynthetic rates [5,6]. Photosynthetic rates decrease mostly due to stomatal closure, but as water stress becomes severe, the inactivation of photosynthetic activity could be due not only to stomatal restrictions, but also to non-stomatal factors related to inhibition of primary photochemistry and electron transport in chloroplasts [7] as well as to the increase in reactive oxygen species (ROS) levels. When the absorbed light energy is not fully used by photosynthesis, it deviates to molecular oxygen, which is abundant in the chloroplasts [8] leading to ROS formation. ROS are highly reactive oxygen species constantly generated by cell organelles as a metabolic by-product; they function as signaling molecules, but their production is spiked upon stress, and plant normal metabolism is seriously disrupted [9].

Plants have a complex antioxidant system to cope with ROS involving enzymes and molecules [10]. The major enzymatic scavengers of ROS are superoxide dismutase (SOD), ascorbate peroxidase (APX), and catalase (CAT) [11]. In addition, plants contain several low-molecular-weight antioxidants, such as ascorbate, glutathione, and phenolic compounds, which are water-soluble, and tocopherol and carotenoids, which are lipid-soluble [12]. Although studies on the enzymatic antioxidant system of olive trees under water deficit have demonstrated that antioxidant enzymes play a major role in protecting olive leaf tissues against oxidative stress [13–15], the role of phenolic compounds on the water stress tolerance of olive trees has received limited attention [5].

As sessile organisms, olive plants have an active secondary metabolism to improve their adaptation to biotic and abiotic stress conditions [16]. The most characteristic secondary metabolites present in olive trees are iridoids, triterpenes, and phenolic compounds. These metabolites accumulate preferentially in leaves and their beneficial effects as antihypertensive for human health due to the coordinated effects of iridoids (oleuropein, oleacein, and ligustroside) and triterpenes (oleanolic acid) have been previously demonstrated [17–19]; their antitumor potential has also been reported [20].

As water stress is a relevant problem, plants have many innate mechanisms that regulate adaptation to stress. Biotechnological attempts to improve adaptations to water deficit with genetic modifications target the overexpression of different genes, as in *Arabidopsis* [21] or cereals, such as rice [22]. In woody plants like the olive tree with such a large cropping surface, genetic modification is not the best choice to improve adaptation. An alternative to genetic modification addressing several targets is the plants' natural associates, soil microorganisms, especially, beneficial strains termed plant growth-promoting rhizobacteria (PGPR).

The term plant growth-promoting rhizobacteria was coined by Kloepper et al. in 1980 [23] to refer to free-living bacteria that inhabit the rhizosphere, which is the soil closely related to the roots. The mechanisms by which these PGPR improve plant fitness have been largely reviewed [24,25]; PGPR affects plant external factors such as nutrient mobilization or biocontrol of soil microorganisms, or alter internal metabolism by affecting endogenous hormonal balance or systemic induction of metabolism at different levels, like photosynthesis or secondary metabolism. Thus, the role of beneficial rhizobacteria to trigger secondary metabolism appears as a promising alternative to increase the levels of bioactive secondary metabolites [26–29], protect against biotic stress [30] and other frequent situations in agriculture [31]. More precisely, protection against salt stress can be enhanced by beneficial rhizobacteria by boosting the ROS-scavenging system, increasing compatible solute concentration, such as proline or soluble sugars, or improving photosynthesis and water use efficiency [6].

Therefore, the use of beneficial rhizobacteria capable of modulating secondary metabolism pathways of plants appears as a biotechnological tool with great potential for this purpose. The application of beneficial strains to improve adaptation to abiotic stress has been widely shown for different species, either woody or herbaceous crops, targeting many mechanisms simultaneously [6] and therefore, with great chances of success. To our knowledge, no studies have been undertaken specifically on olive plants to improve adaptation to salt stress with beneficial rhizobacteria, paying specific attention to bioactive molecules accumulated in leaves; furthermore, if bioactives accumulate in leaves, pruning residues can be transformed into a valuable side product, to obtain enriched extracts with antihypertensive potential. Our rationale is that inoculating the olive trees with beneficial rhizobacteria would simultaneously trigger secondary metabolism pathways as well as other mechanisms also involved in abiotic stress adaptation. We selected 10 bacterial strains from a previous screening in *Pinus* rhizosphere [32] to evaluate their ability to improve olive tree adaptation to salt stress and enhance bioactive contents. To achieve this objective, photosynthesis was measured after 12 months of inoculations on plantlets grown in high saline conditions, photosynthetic pigments, osmolites (soluble sugars, proline), ROS scavenging enzymes (SOD, APX), and antioxidant molecules (phenols, flavonols, and oleouropein) were analyzed as markers of the overall fitness of the plant.

2. Materials and Methods

2.1. Beneficial Strains and Olive Tree Variety

The 10 beneficial strains (L79, L81, L56, L24, L62, L36, G7, L44, K8, and H47) assayed in this study were isolated from the rhizosphere of *Pinus pinea* and *P. pinaster* [32]. They were able to produce siderophores (L79, L81, G7, H47), auxins (L56, L24, L44), auxins and siderophores (L62, L36) or auxins and degrade ACC (K8). Except for L62, a Gram-positive non-esporulated rod, all other strains are Gram-positive esporulated bacilli.

Olea europea (L) var. Arbosana plantlets were used for the study. Plantlets were bought from a commercial producer.

2.2. Inocula Preparation and Delivery to Plants

Bacterial strains were maintained at −80 °C in nutrient broth with 20% glycerol. Inocula were prepared by streaking strains from −80 °C onto plate count agar (PCA) plates, incubating plates at 28 °C for 24 h. Then, they were grown in Luria Broth liquid media (LB) or nutrient broth (only L62) under shaking (1000 rpm.) at 28 °C for 24 h to obtain a 2×10^9 cfu/mL inoculum.

These cultures were adjusted to 1×10^8 cfu/mL and 500 mL were root-inoculated every 15 days from October 2017 to October 2018.

2.3. Experimental Design

Six-month olive plantlets were transplanted into 5 L pots with soil from the Guadalquivir Marshes. Plants were arranged in lines on an experimental plot within the marshes (37°06′34.5′′ N, 6°20′22.7′′ W); pot position was changed every two weeks to avoid side-effects. Plants were watered every 15 days. The electric conductivity of water was 8.20 dS/m and of soil it was 6.07 dS/m.

Bacteria were root-inoculated by soil drench every 15 days from October 2017 to October 2018; so plants received 500 mL of water every week, alternating inoculum and water. Six plants per treatment were inoculated, being one bacterial strain a treatment. Samples were taken in October 2018 and photosynthesis was measured (fluorescence and CO_2 fixation). Leaves were powdered in liquid nitrogen and stored at −60 °C till analysis. Photosynthetic pigments were determined as well as their antioxidant capacity, analyzing both the enzymatic and non-enzymatic apparatus. Superoxide dismutase (SOD) and ascorbate peroxidase (APX) activities were determined as indicators of the enzyme apparatus, and total phenols, flavonols and oleuropein, as indicators of the non-enzymatic pool. The osmoprotective effect was evaluated by analyzing compatible solutes (proline and soluble sugars).

2.4. Photosynthesis (Chlorophyll Fluorescence)

Photosynthetic efficiency was determined through the chlorophyll fluorescence emitted by photosystem II. Chlorophyll fluorescence was measured with a pulse amplitude modulated (PAM) fluorometer (Hansatech FM2, Hansatech, Inc., UK). After dark-adaptation of leaves, the minimal fluorescence (Fo; dark-adapted minimum fluorescence) was measured with a weak modulated irradiation (1 μmol m^{-2} s^{-1}). Maximum fluorescence (Fm) was determined for the dark-adapted state by applying a 700 ms saturating flash (9000 μmol m^{-2} s^{-1}). The variable fluorescence (Fv) was calculated as the difference between the maximum fluorescence (Fm) and the minimum fluorescence (Fo). The maximum photosynthetic efficiency of photosystem II (maximal PSII quantum yield) was calculated as Fv/Fm. Immediately, the leaf was continuously irradiated with red-blue actinic beams (80 μmol m^{-2} s^{-1}) and equilibrated for 15 s to record Fs (steady-state fluorescence signal). Following this, another saturation flash (9000 μmol m^{-2} s^{-1}) was applied and then Fm' (maximum fluorescence under light-adapted conditions) was determined. Other fluorescent parameters were calculated as follows: the effective PSII quantum yield ΦPSII = (Fm' − Fs)/Fm' [33]; and the non-photochemical quenching coefficient NPQ = (Fm − Fm')/Fm'. All measurements were carried out in the 6 plants of each treatment.

2.5. Photosynthesis (CO$_2$ Fixation)

Net photosynthetic rate, (Pn) (mmol CO$_2$/m^2), transpiration rate, E (mmol/m^2 s) and stomatal conductance, C (mmol/m^2 s) were measured with a portable photosynthetic open-system (CI-340, CID, Camas, WA, USA) [34].

Water use efficiency (WUE) was calculated as net photosynthesis (Pn) divided by transpiration (E) as an indicator of stomatal efficiency to maximize photosynthesis minimizing water loss due to transpiration.

2.6. Photosynthetic Pigments: Chlorophylls and Carotenoids

Extraction was done according to [35]. One hundred mg of leaves powdered in liquid nitrogen was dissolved in 1 mL of acetone 80% (*v/v*), incubated overnight at 4 °C and then centrifuged 5 min at 10,000 rpm in a Hermle Z233 M-2 centrifuge. One mL of acetone 80% was added to the supernantant and was mixed with a vortex. Immediately afterward, absorbance at 647, 663, and 470 nm was measured on a Biomate 5 spectrophotometer to calculate chlorophyll a, chlorophyll b, and carotenoids (xanthophylls + carotenes) using the formulas indicated below [35,36].

$$\text{Chl a } (\mu g/g \text{ FW}) = [(12.25 \times \text{Abs}_{663}) - (2.55 \times \text{Abs}_{647})] \times V(\text{mL})/\text{weight (g)}.$$

$$\text{Chl b } (\mu g/g \text{ FW}) = [(20.31 \times \text{Abs}_{647}) - (4.91 \times \text{Abs}_{663})] \times V(\text{mL})/\text{weight (g)}.$$

$$\text{Carotenoids } (\mu g/g \text{ FW}) = [((1000 \times \text{Abs}_{470}) - (1.82 \times \text{Chl a}) - (85.02 \times \text{Chl b}))/198] \times V(\text{mL})/\text{weight (g)}.$$

Tubes were protected from light throughout the whole process.

2.7. Enzymatic Antioxidants: Superoxide Dismutase (SOD) and Ascorbate Peroxidase (APX)

Before assessing enzymatic activities, soluble proteins were extracted by suspending 100 mg of powder in 1 mL of potassium phosphate buffer 0.1 M, pH 7.0, containing 2 mM phenylmethylsulfonyl fluoride (PMSF). After sonication for 10 min and centrifugation for 10 min at 14,000 rpm, the supernatant was aliquoted, frozen in liquid nitrogen and stored at −80 °C for further analysis of APX, SOD, and proteins. All the above operations were carried out at 0–4 °C.

To measure the amount of total protein from plant extract, 250 μL of Bradford reagent and 5 μL of sample and BSA dilutions were inoculated in ELISA 96 well plates and incubated for 30 min at room temperature and measured using a plate reader at an absorbance of 595 nm. A calibration curve was constructed from commercial BSA dilutions expressed in milligrams. The protein units were expressed as mg/μL.

APX was measured by the method of Garcia-Limones [37]. The reaction mixture consisted of 50 mM potassium phosphate buffer, pH 7.0, 0.25 mM sodium ascorbate, 5 mM H_2O_2 and 100 μL of enzyme extract in a final volume of 1.2 mL. Adding H_2O_2 started the reaction and the oxidation of ascorbate was determined by the decrease in A290. The extinction coefficient of 2.8 mM^{-1} cm^{-1} was used to calculate activity. One unit of APX activity is defined as the amount of enzyme that oxidizes 1 mmol min^{-1} of ascorbate under the above assay conditions.

SOD activity was determined following the specifications of the SOD activity detection kit (SOD Assay Kit-WST, Sigma-Aldrich, Darmstadt, Germany). With this method, xanthine is converted to superoxide radical ions, uric acid, and hydrogen peroxide by xanthine oxidase (XO). Superoxide reacts with WST1 to generate a product that absorbs at around 440 nm. SOD prevents the reduction of WST1 to WST-1formazan, thus reducing the absorption at 440 nm, which is proportional to SOD activity; the rate of the reduction of WST1 with O_2 is linearly related to the xanthine oxidase (XO) activity. The unit used for this activity was: % inhibition of WST reduction.

2.8. Osmolites: Proline and Soluble Sugars

An ethanolic extraction was prepared from 0.25 g of powder and 5 mL of ethanol 70% (*v/v*) incubated at 100 °C for 20 min.

For proline determination 1 mL of ninhydrin reagent freshly prepared (1 g of ninhydrin dissolved in 60 mL of acetic acid, 20 mL of ethanol and 20 mL of water) was mixed with 0.5 mL of the plant ethanol extract and heated at 95 °C for 20 min. Finally, absorbance at 520 nm was measured. Results are expressed as μmol/g.

A soluble sugars determination was performed following Yemm and Willis [38]. Briefly, the following reaction was prepared: 3 mL of the reactive (200 mg of antrone + 100 mL of 72% sulfuric acid) and 0.1 mL of the plant ethanol extract. The reaction was incubated in a bath at 100 °C for 10 min. Once it was cold, absorbance was measured at 620 nm. To calculate soluble sugar concentration the following equation was used:

$$\mu g/g = [(Abs_{620} - 0.016)/0.02]/\text{weight (g)}/1000$$

2.9. Total Phenols and Flavonols

Leaf extracts were prepared from 0.25 g of leaves (powdered in liquid nitrogen) in 2.25 mL methanol 80%, sonicated for 10 min and centrifuged for 5 min at 5000 rpm.

Total phenols were quantitatively determined with Folin-Ciocalteu agent (Sigma. Aldrich, St Louis, MO, USA) by a colorimetric method described by Xu and Chang [39], with some modifications, gallic acid was used as standard (Sigma-Aldrich, St Louis, MO, USA). Twenty microlitres of extract were mixed with 0.250 mL of Folin-Ciocalteu 2 N and 0.75 mL of Na_2CO_3 20% solution. After 30 min at room temperature, absorbance was measured at 760 nm. A gallic acid calibration curve was made (r = 0.99). Results are expressed in mg of gallic acid equivalents per 100 g of fresh weight (FW).

Total flavonols were quantitatively determined through the test described by Jia et al. [40], using catechin as standard (Sigma-Aldrich, St Louis, MO, USA). One milliliter of the extract was added to a flask of 10 mL with 4 mL of distilled water. After that 0.3 mL of $NaNO_2$ 5%, and 0.3 mL of $AlCl_3$ 10% were added after 5 min. One minute later, 2 mL of NaOH 1 M were added, and distilled water was added util 10 mL of total volume. The solution was mixed and measured at 510 nm. A catechin calibration curve was made (r = 0.99). Results are expressed as mg of catechin equivalents per 100 g of fresh weigh (FW).

2.10. Oleuropein Extraction and TLC Analysis

Oleuropein was determined according to the European Pharmacopoeia. One gram of the powdered samples was extracted with 10 mL of methanol under reflux for 15 min. After cooling, samples were filtered and 10 µL was loaded as a band on a TLC silica gel plate; the reference solution contained 10 mg of oleuropein and 1 mg of rutoside trihydrate in 1 mL of methanol. Plates were incubated on a chromatography tank and allowed to develop over a path of 10 cm, being the mobile phase water/methanol/methylene chloride (1.5:15:85 *v/v/v*). Plates were dried in air. Detection of oleuropein was done by spraying with vanillin sulphuric acid reagent after followed by heating for 5 min at 100–105 °C; the brownish-green zone appeared in the middle of the plate was oleuropein and a brownish-yellow zone near the application point was rutoside.

Quantification was done with the image analysis program Quantity One v4.6.8 (Biorad, CA, USA), based on the density and concentration of the oleuropein spot from the reference sample.

2.11. Statistics

A Principal Components Analysis (PCA) with all the parameters measured for the ten strains was performed with Canoco^{TM} for Windows v.4.5 software (Microcomputer power, Ithaca, USA) [41]. Scaling was performed withinter-species correlation and was achieved dividing by the standard deviation.

To evaluate treatment effects, one way ANOVA (Statpgraphcis Centurion XVIII) were performed for each of the variables. When significant differences appeared ($p < 0.05$), the LSD test (least significant difference) from Fisher was used.

3. Results

Three main groups can be defined in the ordination provided by principal component analysis (PCA) (Figure 1). The group in the upper part of axis I (dotted line) includes bacteria L36, K8, L44, and L81 and is mainly influenced by photosynthetic pigments concentration and, secondarily, by APX and SOD activities as shown by the length of the vectors. A second group including K8, L44, L81, H47, and L56 (black dashed line) on the left of axis I, can be defined based on phenols and flavonols concentration. A third group formed by L56, L24, and L62 (grey dashed line), is determined by oleuropein concentration and water use efficiency (WUE).

All assayed strains modified photosynthetic parameters (Figure 2), however, Fv/Fm was within normal values (0.82–0.85) and was not significantly affected by any strain. L81, L24, and K8 significantly increased photochemical quenching and all strains except L79 and L36 decreased energy dissipation (NPQ). All of them significantly increased net photosynthesis in terms of CO_2 fixation, with an outstanding performance of K8 and L24 that caused 3-fold increases, while all others were in the range of 2-fold increases. Water use efficiency, calculated as the value of net photosynthesis divided by the transpiration rate, was also significantly increased by all strains, with an outstanding performance of L62 that caused a 5-fold increase on WUE, and L24 and L44 in second place, in the range of a 3.5 increase on WUE (Figure 3).

As regards to photosynthetic pigments (Figure 4), controls had around 63 mg/g chlorophyll a, 29.69 chlorophyll b, and 72.3 mg/g carotenes. Most strains maintained or decreased the amount of chlorophylls, except for strain L36 that increased chlorophylls and K8 that increased chlorophyll a; none modified the chlorophyll a/b ratio. The general trend was to lower carotene concentration, except for L36 and L44.

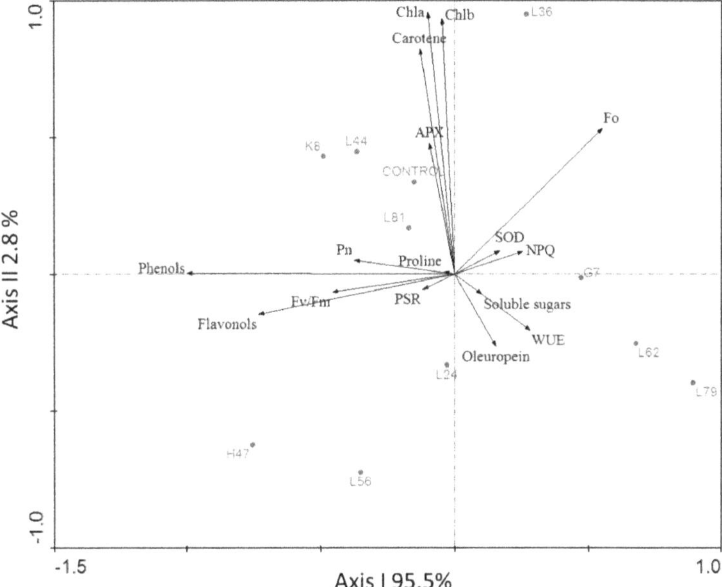

Figure 1. Ordination provided by the principal components analysis (PCA) performed with all the parameters measured for the ten strains. APX, ascorbate peroxidase; SOD, superoxide dismutase; NPQ, non-photochemical quenching coefficient; WUE, Water Use Efficiency. Percentages in the axis correspond to the variance absorbed by each of these two first axes.

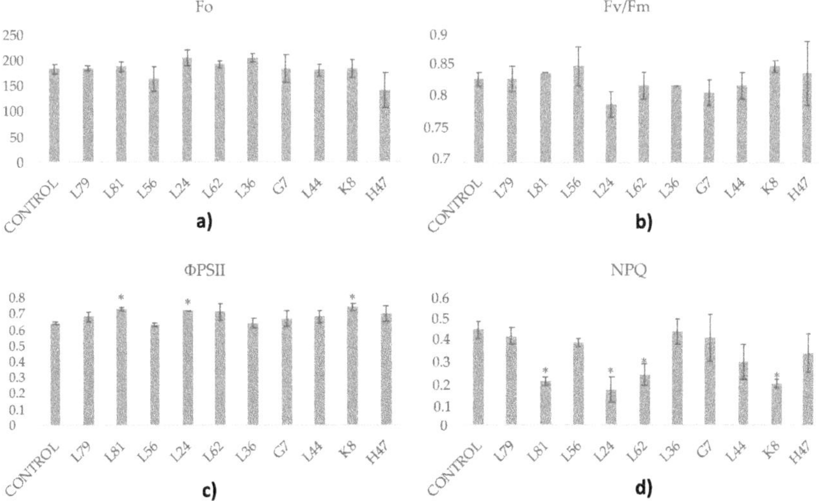

Figure 2. Photosynthetic parameters related to photosystems and light reactions. (**a**) Minimal fluorescence after 20-min dark-adaptation (Fo). (**b**) Maximal PSII quantum yield (Fv/Fm), (**c**) effective PSII quantum yield (ΦPSII) and (**d**) non-photochemical quenching coefficient (NPQ) measured in olive tree plants treated with the ten strains and the non-inoculated control. For each treatment and parameter average value ± standard error value ($n = 6$) is presented. Asterisks (*) represent significant differences with the control according to the LSD test ($p < 0.05$).

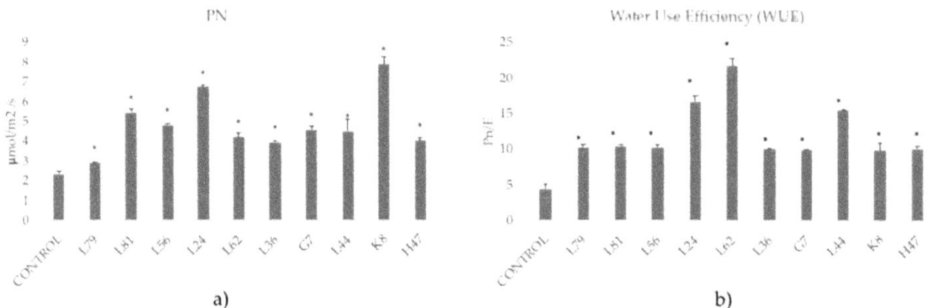

a)   b)

Figure 3. Photosynthetic parameters related to C fixation measured in olive tree plants treated with the ten strains. (**a**) Net photosynthesis (PN) measured as the CO_2 fixed by the leaves (μmol CO_2 /m^2 s). (**b**) Water Use Efficiency (WUE) calculated as PN divided by transpiration rate (μmolH_2O /m^2 s). Average values of the replicates with standard error bars are represented ($n = 6$). Asterisks (*) represent significant differences with the control according to the LSD test ($p < 0.05$).

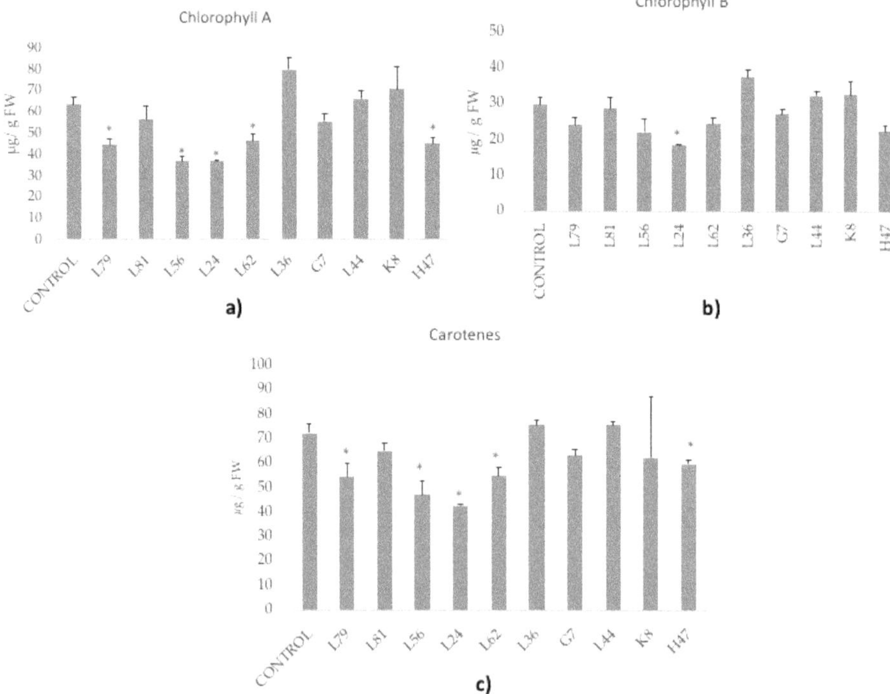

Figure 4. Photosynthetic pigments concentration (μg/g FW). (**a**) Chlorophyll a, (**b**) Chlorophyll b and (**c**) Carotenoids were measured in olive tree leaves treated with the ten strains. For each treatment and parameter average value ± standard error value is presented ($n = 6$). Asterisks (*) represent significant differences with the control according to the LSD test ($p < 0.05$).

As for the enzymatic antioxidant systems (Figure 5), only L81 increased the activity of SOD and L44 that of APX, while no changes or slight decreases were induced by the other 8 strains.

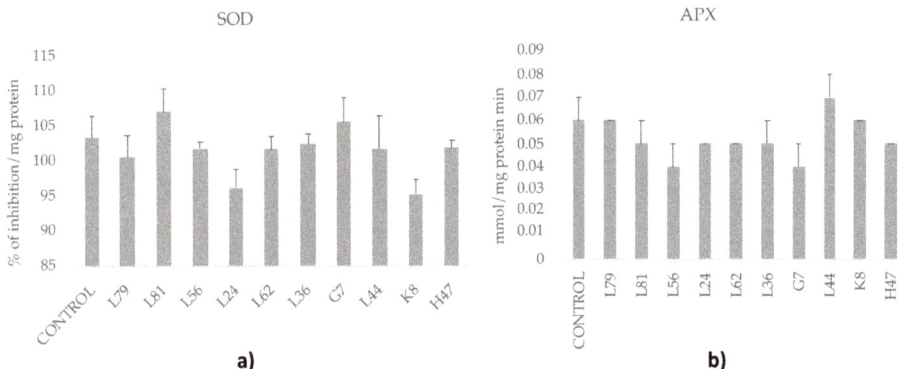

Figure 5. Enzyme activities related to oxidative stress. (**a**) Superoxide dismutase (SOD), (**b**) Ascorbate peroxidase (APX) activities measured in olive tree leaves treated with the ten strains. Enzymatic activities were calculated as mmol/mg protein min (for APX) and % of inhibition/mg protein (for SOD). For each treatment and parameter average value ± standard error value is presented ($n = 6$). There are no significant differences according to the LSD test ($p < 0.05$).

Soluble sugars in controls were 3.77 mg/g and proline contents 0.37 µmol/g (Figure 6). Soluble sugars contents were significantly increased by all treatments, ranging from 10% (L24) to 30% (L62 (Figure 6a)). Similarly to soluble sugars, proline was significantly increased by most strains ranging from 20% (L56) to 50% (G7, H47); only L81 and L62 showed similar proline contents than controls (Figure 6b).

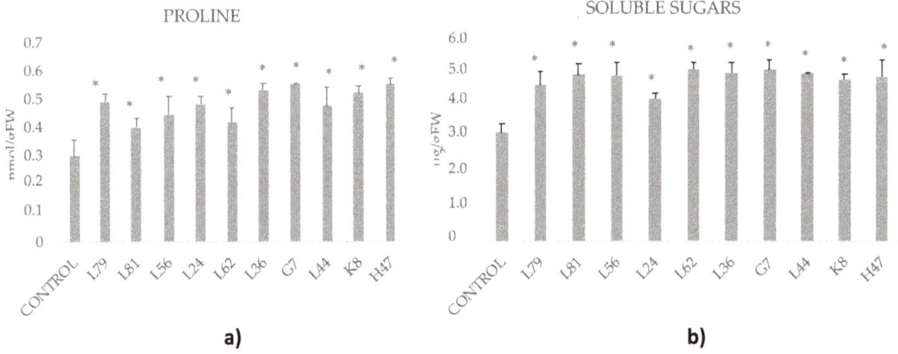

Figure 6. (**a**) Proline (nmol/g fresh weight) and (**b**) soluble sugars concentration (µmol/g fresh weight) measured in olive tree leaves treated with the ten strains. Average values of the replicates with standard error bars are represented ($n = 6$). Asterisks (*) represent significant differences with the control according to the LSD test ($p < 0.05$).

As for bioactives (Figure 7), controls had high phenols concentration (1295 meq gallic acid/100 g FW) and low flavonols (6.46 meq catechin/100 g FW), and 9.11 mg oleuropein/g. Strains L81 and L56 increased the concentration of phenols and flavonols in leaves and L44, K8, and H47 only that of phenols. Only L81 increased oleuropein concentration (12%), L62 did not affect it and most strains decreased the amount of oleuropein, a group with minor decreases (L56, L24, L79) and another group with major decrease.

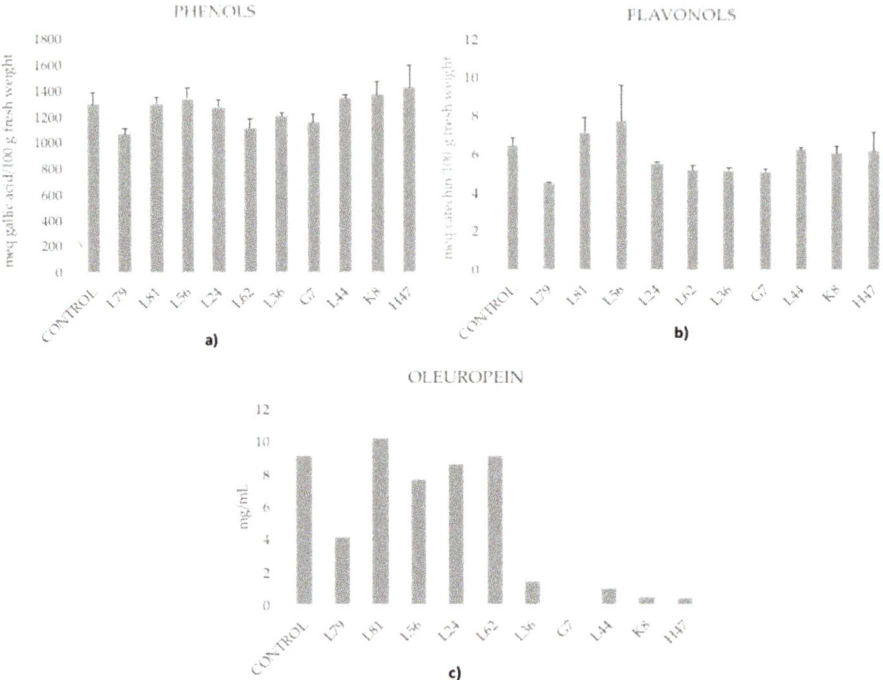

Figure 7. Bioactives. (**a**) Phenols (meq gallic acid/100 g fresh weight), (**b**) flavonols (meq catechin/100 g fresh weight) and (**c**) oleuropein (mg/mL) concentration measured in olive tree leaves treated with the ten strains. For each treatment and parameter average value ± standard error value is presented. There are no significant differences according to the LSD test ($p < 0.05$).

4. Discussion

In this study, the ability of the ten beneficial rhizobacteria assayed to modify plant physiology of olive plantlets growing in pots with high electric conductivity of soil and water when root-inoculated has been evidenced. According to Chartzouaskis [42], values of water (8.20 dS/m) and soil (6.07 dS/m) electric conductivity, indicate severe salinity for olive trees in the present study. In these harsh conditions, all strains are able to trigger plant adaptative metabolism, improving net photosynthesis and mainly affecting osmolite concentration. The positive effect of these strains was expected since the original screening rendered several beneficial strains, two of which (L62 and L81) are also tested in this experiment [30,32].

In general, the effects of bacteria on plant metabolism and physiology target photosynthetic pigments, photosynthesis, and osmolites [8] as revealed by the principal components analysis (PCA), a multivariate analysis that provides an ordination of the samples based on their similarity considering all the variables analyzed. In this ordination (Figure 1), samples are grouped mainly due to the differential effects of bacteria in those three variables. The group in the upper part of axis I (dotted line) includes bacteria that increase (L36 and K8) or maintain (L44 and L81) photosynthetic pigments. The position of these two bacteria (L44 and L81) in the ordination is also determined by the effect they have in APX and SOD, respectively. All the other bacteria decreased pigment concentration. Bacteria in the second group (black dashed line), K8, L44, L81, H47, and L56, increase leaf phenol concentrations, and L81 and L56 also increase flavonols. A third group formed by L56, L24, and L62 (grey dashed line) includes the three bacteria that maintain oleuropein concentration similar to control plants, while the other bacteria reduce this concentration, except for L81, the only strain that increases oleuropein concentration in leaves. Finally, L24 and L62 induced the highest water use efficiency in the plants. Therefore, all these parameters are differentially affected by the different strains while proline, soluble sugars concentration, and net photosynthetic rate are similarly triggered by all bacteria [6], as can be noticed by the shorter length of vectors.

Under mild and moderate water stress, photosynthetic rate decreases in olive plants mostly due to stomatal closure [43]. However, as water stress becomes severe, the inactivation of photosynthetic activity could be due not only to stomatal closure but also to non-stomatal factors related to inhibition of primary photochemistry and electron transport in chloroplasts [7]. A decrease in chlorophylls under salt stress has been explained by pigment destruction after the ROS peak [9]. However, if this was the case, photosynthesis would be impaired due to cell membrane alterations and lack of pigments. Interestingly, our data show increased effective PSII quantum yield (ΦPSII) (L81, L24 and K8), and lower NPQ, suggesting that bacteria are triggering an innate plant protective mechanism against the excess of light entering the system, and making better use of the energy fixed.

Under salt stress conditions olive leaves become thicker [44] compromising CO_2 diffusion to chloroplasts [45,46]. Photosynthesis is reduced under saline stress in olive trees [44,47,48], but with different effects on the CO_2 assimilation rate depending on salt concentration. Interestingly, all strains increased net photosynthesis, consistent with the reported modification of carbohydrate production and sink utilization that leads to downregulate feedback photoinhibition and boost plant energy metabolism [49], probably providing C scaffoldings for secondary metabolites and osmolyte synthesis and accumulation. Despite the significant increase in C fixation induced by all strains, water use efficiency (WUE) was different, with a striking two-fold increase for most strains except for L62, which showed a 4-fold increase, and L44 and L24, showing a 3-fold increase (Figure 3). This indicates a strong improvement in plant fitness in a high saline environment, suggesting activation of protective systems and highlighting the different mechanisms involved in adaptation to harsh conditions and supporting the multivariate solution provided by PGPR [6].

There is a demonstrated relationship between compatible solutes and photosynthesis. All strains increased concentration of compatible solutes as a protective mechanism, since they sequester water molecules, protect cell membranes and protein complexes, and allow the metabolic machinery to continue functioning [8]. Consistent with our data, carbohydrates are the most common solutes accumulated in olive tree tissue under water deficit conditions [6,50], and all strains significantly increased them, being thus a primary defense mechanism [49]. The close relationship between net photosynthetic rate and proline content reported by BenAhmed et al. [13], is consistent with our data, as proline was increased by all except for L81 and L62, confirming the important role of this osmolyte in the maintenance of photosynthetic activity and plant homeostasis. The different behaviors suggest the involvement of other factors, as L81 and L62 performed outstandingly.

Bacterial strains modify differently innate plant mechanisms to cope with ROS [29,51]; while the enzyme ROS scavenging system is hardly modified by bacteria, phenolic compounds are. The enzyme system is probably in its maximum natural activation, which cannot be further increased by bacteria; however, L81 is still able to significantly enhance SOD activity. Although studies on the enzymatic antioxidant system of the olive tree under water deficit have demonstrated that antioxidant enzymes play a major role in protecting olive leaf tissue against oxidative stress [13–15], limited attention has been given to the effect of phenolic compounds on water stress tolerance. Phenolic compounds are constitutively present in all higher plants. However, phenylpropanoid metabolism is often induced when plants are exposed to a wide range of environmental stresses [52], including bacteria [26,53]. In view of our results, abiotic stress as well as beneficial rhizobacteria modified the antioxidant pools of the plants but in an uncoupled way. Some of the assayed strains increased total phenols and flavonols (L81 and K8), and oleuropein (only L81), while another group decreased concentrations of these metabolites (L79, L24, L62, and G7). Among the first group, L81 increased SOD enzyme activity. From the second group, L79 increased SOD and L24 decreased both SOD and APX enzymatic activities, which suggests either higher oxidative stress as a consequence or other mechanisms to cope with

ROS (Figure 5). The different influence of bacterial strains on ROS scavenging enzyme activity has been described before to be strain-specific [28]. García-Cristobal et al [28] reported the ability of a *Chryseobacterium* strain to enhance ROS scavenging enzymes activity in salt and pathogen stressed rice plants, while a *Pseudomonas* strain enhanced protection by increasing defensive enzymes, not ROS scavenging enzymes.

Under salt stress conditions, phenolic compounds produced in leaves increase [5,6,15]. However, in this work only five strains increased total phenolics concentration and only two significantly increased total flavonols, while others decreased them, reinforcing the species specificity between plant and bacteria [28,53], a receptor-mediated effect, hence highly specific. Irrespective of the final phenolics balance, all bacteria have altered this pathway, confirming the role of this pathway in adaptation; not only phenolics behave as antioxidants, but other derived molecules may also have this role [27,54]. Considering all this data together, it seems that bacteria are lowering photosynthetic pigments concentration, suggesting this effect as a mechanism to decrease oxidative stress due to photosynthesis, especially since the enzymatic antioxidant pool is not affected or even decreased, and bioactives are only increased by half of the strains. Finally, oleuropein, a bioactive molecule accumulated in leaves [18], with a proposed role as a protective molecule against biotic stress due to its potential as a cross-linking agent [54], is only increased under the influence of L81. Again, two strategies are depicted, either slightly lowering its concentration, or minimizing it reinforcing the hypothesis of each strain activating different mechanisms of plant adaptation.

Not only olive oil is obtained from olive trees; also solid residues obtained in considerable amounts during olive oil production and elaboration of table olives are of great concern in the Mediterranean area, as these by-products accumulate in large amounts. Great progress has been made to recycle these residues obtaining an economic profit, like obtaining activated carbon [55–58] or fuel for the generation of heat and electricity [59–61]. However, olive leaves, which are produced in large amounts, render scarce profits at present. Nevertheless, the market for natural ingredients and additives is rapidly growing, with such products obtaining high prices. Increased concentrations of phenolic compounds and especially oleuropein, with strong antihypertensive potential, reinforces the potential of olive leaves in the field of a circular economy. Furthermore, delivering beneficial strains to edible plants has improved beneficial effects for health as not only the targeted metabolites are increased, but there is a general physiological change that results in improved effects on health [53].

In summary, delivering beneficial strains improves adaptation to high saline conditions, mainly affecting osmolytes and improving net photosynthesis and water use efficiency. Interestingly, L81 differentially increases oleuropein constituting a good treatment to improve the potential of olive leaves for its antihypertensive effects. L62 is the one to improve WUE, which is especially good to improve plant adaptation to harsh conditions of low water availability.

5. Conclusions

In view of these data, it is evidenced that all bacterial strains improve plant adaptation increasing osmoprotection and net photosynthesis but they differentially affect the enzymatic and non-enzymatic antioxidant systems. Bacteria able to increase bioactive concentration and therefore potential benefits of olive leaves on health may also contribute to a circular economy, recycling pruning residues.

Author Contributions: Conceptualization, F.J.G.-M., B.R.-S., A.G.-V.; methodology, A.G.-V., E.G.-C.; formal analysis, E.G.-C., M.B.M.-P.; investigation, E.G.-C.; resources, F.J.G.-M., B.R.-S.; data curation, A.G.-V., E.G.-C.; writing—original draft preparation, A.G.-V.; writing—review and editing, A.G.-V., B.R.-S.; supervision, F.J.G.-M., B.R.-S. All authors have read and agreed to the published version of the manuscript.

Funding: This research received no external funding.

Acknowledgments: Authors would like to thank J.C. for providing experimental fields and local help.

Conflicts of Interest: The authors declare no conflict of interest.

References

1. Vossen, P. Olive Oil: History, Production, and Characteristics of the World's Classic Oils. *Hortscience* **2007**, *42*, 1093–10100. [CrossRef]
2. Tous, J.; Romero, A.; Hermoso, J.J. New trends in olive orchard design for continuous mechanical harvesting. *Adv. Hort. Sci.* **2010**, *24*, 43–52.
3. Bacelar, E.A.; Santos, D.L.; Moutinho-Pereira, J.M.; Gonçalves, B.C.; Ferreira, H.F.; Correia, C.M. Immediate responses and adaptive strategies of three olive cultivars under contrasting water availability regimes: Changes on structure and chemical composition of foliage and oxidative damage. *Plant Sci.* **2006**, *170*, 596–605. [CrossRef]
4. Mousavi, S.; Regni, L.; Bocchini, M.; Mariotti, R.; Cultrera, N.G.M.; Mancuso, S.; Googlani, J.; Chakerolhosseini, M.R.; Guerrero, C.; Albertini, E.; et al. Physiological, epigenetic and genetic regulation in some olive cultivars under salt stress. *Sci. Rep.* **2019**, *9*, 1093. [CrossRef]
5. Petridis, A.; Therios, I.; Samouris, G.; Tananaki, C. Salinity-induced changes in phenolic compounds in leaves and roots of four olive cultivars (*Olea europaea* L.) and their relationship to antioxidant activity. *Environ. Exp. Bot.* **2012**, *79*, 37–43. [CrossRef]
6. Ilangumaran, G.; Smith, D.L. Plant growth promoting rhizobacteria in amelioration of salinity stress: A systems biology perspective. *Front. Plant Sci.* **2017**, *8*, 1768. [CrossRef]
7. Boyer, J.S.; Armond, P.A.; Sharp, P.E. Light stress and leaf water relations. In *Photoinhibition, Topics in Photo-Synthesis*; Kyle, D.J., Osmond, C.D., Arntzen, C.J., Eds.; Elsevier: Amsterdam, The Netherlands, 1987; Volume 9, pp. 111–122.
8. Chaves, M.M.; Pereira, J.S.; Maroco, J. Understanding plant response to drought-from genes to the whole plant. *Funct. Plant Biol.* **2003**, *30*, 1–26. [CrossRef]
9. Apel, K.; Hirth, H. Reactive oxygen species: Metabolism, oxidative stress and signal transduction. *Ann. Rev. Plant Biol.* **2004**, *55*, 373–399. [CrossRef]
10. Gill, S.S.; Tuteja, N. Reactive oxygen species and antioxidant machinery in abiotic stress tolerance in crop plants. *Plant Physiol. Biochem.* **2010**, *48*, 909–930. [CrossRef]
11. Noctor, G.; Foyer, C.H. Ascorbate and glutathione: Keeping active oxygen under control. *Ann. Rev. Plant Physiol. Plant Mol. Biol.* **1998**, *49*, 249–279. [CrossRef]
12. Grace, S.C. Phenolics as antioxidants. In *Antioxidants and Reactive Oxygen Species in Plants*; Smirnoff, N., Ed.; Blackwell Publishing Ltd.: Oxford, UK, 2005; pp. 141–168.
13. BenAhmed, C.; BenRouina, B.; Sensoy, S.; Boukhris, M.; BenAbdallah, F. Changes in gas exchange, proline accumulation and antioxidative enzyme activities in three olive cultivars under contrasting water availability regimes. *Environ. Exp. Bot.* **2009**, *67*, 345–352. [CrossRef]
14. Sofo, A.; Dichio, B.; Xiloyannis, C.; Masia, A. Effects of different irradiance levels on some antioxidant enzymes and on malondialdehyde content during rewatering in olive tree. *Plant Sci.* **2004**, *166*, 293–302. [CrossRef]
15. Ennajeh, M.; Vadel, A.M.; Khemira, H. Osmoregulation and osmoprotection in the leaf cells of two olive cultivars subjected to severe water deficit. *Acta Physiol. Plant.* **2009**, *31*, 711–721. [CrossRef]
16. Oh, M.; Trick, H.N.; Rajashekar, C.B. Secondary metabolism and antioxidants are involved in environmental adaptation and stress tolerance in lettuce. *J. Plant Physiol.* **2009**, *166*, 180–191. [CrossRef]
17. El Riachy, M.; Priego-Capote, F.; León, L.; Rallo, L.; Luque de Castro, M.L. Hydrophilic antioxidants of virgin olive oil. Part 1: Hydrophilic phenols: A key factor for virgin olive oil quality. *Eur. J. Lipid Sci. Technol.* **2011**, *113*, 678–691. [CrossRef]
18. Bulotta, S.; Oliverio, M.; Russo, D.; Procopio, A. Biological activity of oleuropein and its derivatives. In *Natural Products*; Ramawat, K.G., Mérillon, J.M., Eds.; Springer: Berlin/Heidelberg, Germany, 2013; pp. 3605–3638.
19. Alagna, F.; Mariotti, R.; Panara, F.; Caporali, S.; Urbani, S.; Veneziani, G.; Esposto, S.; Taticchi, A.; Rosati, A.; Rao, R.; et al. Olive phenolic compounds: Metabolic and transcriptional profiling during fruit development. *BMC Plant Biol.* **2012**, *12*, 162. [CrossRef]
20. Celano, M.; Maggisano, V.; Lepore, S.M.; Russo, D.; Bulotta, S. Secoiridoids of olive and derivatives as potential coadjuvant drugs in cancer: A critical analysis of experimental studies. *Pharmacol. Res.* **2019**, *142*, 77–86. [CrossRef]

21. Pehlivan, N.; Sun, L.; Jarrett, P.; Yang, X.; Mishra, N.; Chen, L.; Kadioglu, A.; Shen, G.; Zhang, H Co-overexpressing a plasma membrane and a vacuolar membrane Sodium/Proton antiporter significantly improves salt tolerance in transgenic Arabidopsis plants. *Plant Cell Physiol.* **2016**, *57*, 1069–1084. [CrossRef]

22. Yasmin, F.; Biswas, S.; Jewel, G.; Elias, S.; Seraj, Z. Constitutive overexpression of the plasma membrane Na$^+$/H$^+$ antiporter for conferring salinity tolerance in rice. *Plant Tissue Cult. Biotechnol.* **2016**, *25*, 257–272. [CrossRef]

23. Kloepper, J.W.; Schroth, M.N.; Miller, T.D. Effects of rhizosphere colonization by plant growth-promoting rhizobacteria on potato plant development and yield. *Phytopathology* **1980**, *70*, 1078–1082. [CrossRef]

24. Sayyed, R.Z.; Reddy, M.S.; Al-Turki, A.I. *Recent Trends in PGPR Research for Sustainable Crop Productivity*; Scientific Publishers: Jodhpur, India, 2016; p. 258. ISBN 978-81-7233-990-6.

25. Rosier, A.; Medeiros, F.H.V.; Bais, H.P. Defining plant growth promoting rhizobacteria molecular and biochemical networks in beneficial plant-microbe interactions. *Plant Soil.* **2018**, *428*, 35–55. [CrossRef]

26. Garcia-Seco, D.; Zhang, Y.; Gutierrez-Mañero, F.J.; Martin, C.; Ramos-Solano, B. Application of Pseudomonas fluorescens to Blackberry under Field Conditions Improves Fruit Quality by Modifying. Flavonoid Metabolism. *PLoS ONE.* **2015**, *10*, e0142639. [CrossRef] [PubMed]

27. Algar, E.; Ramos-Solano, B.; García-Villaraco, A.; Saco Sierra, M.D.; Martín Gómez, M.S. Gutiérrez-Mañero, F.J. Bacterial Bioeffectors Modify Bioactive Profile and Increase Isoflavone Content in Soybean Sprouts (Glycine max var Osumi). *Plant Foods Hum. Nutr.* **2013**, *68*, 299–305. [CrossRef] [PubMed]

28. Garcia-cristobal, J.; García-Villaraco, A.; Ramos, B.; Gutierrez-Mañero, J.; Lucas, J.A. Priming of pathogenesis related-proteins and enzymes related tooxidative stress by plant growth promoting rhizobacteria on riceplants upon abiotic and biotic stress challenge. *J. Plant Physiol.* **2015**, *188*, 72–79. [CrossRef]

29. Lucas, J.A.; García-Cristobal, J.; Bonilla, A.; Ramos, B.; Gutierrez-Mañero, J. Beneficial rhizobacteria from rice rhizosphere confers high protection against biotic and abiotic stress inducing systemic resistance in rice seedlings. *Plant Physiol. Biochem.* **2014**, *82*, 44–53. [CrossRef]

30. Barriuso, J.; Ramos Solano, B.; Gutiérrez Mañero, F.J. Protection Against Pathogen and Salt Stress by Four Plant Growth-Promoting Rhizobacteria Isolated from Pinus sp. on Arabidopsis thaliana. *Phytopathology* **2008**, *98*, 666–672. [CrossRef]

31. Gutiérrez-Mañero, F.J.; García-Villaraco, A.; Lucas, J.A.; Gutiérrez, E.; Ramos-Solano, B. Inoculant/Elicitation Technology to Improve Bioactive/Phytoalexin Contents in Functional Foods. *Int. J. Curr. Microbiol. App. Sci.* **2015**, *4*, 224–241.

32. Barriuso, J.; Pereyra, M.T.; García, J.A.L.; Megías, M.; Gutierrez-Mañero, F.J.; Ramos-Solano, B. Screening for putative PGPR to improve establishment of the symbiosis *Lactarius deliciosus-Pinus sp.*. *Microb. Ecol.* **2005**, *50*, 82–89. [CrossRef]

33. Genty, B.; Briantais, J.M.; Baker, N.R. The relationship between the quantum yield of photosynthetic electron transport and quenching of chlorophyll fluorescence. *BBA-Gen. Subj.* **1989**, *990*, 87–92. [CrossRef]

34. Schlosser, A.J.; Martin, J.M.; Hannah, L.C.; Giroux, M.J. The maize leaf starch mutation has diminished field growth and productivity. *Crop Sci.* **2012**, *52*, 700–706. [CrossRef]

35. Porra, R.J.; Thompson, W.A.; Kriedemann, P.E. Determination of accurate extinction coefficients and simultaneous equations for assaying chlorophylls a and b extracted with four different solvents: Verification of the concentration of chlorophyll standards by atomic absorption spectroscopy. *BBA Bioenerg.* **1989**, *975*, 384–394. [CrossRef]

36. Lichtenthaler, H.K. Chlorophylls and carotenoids: Pigments of photosynthetic biomembranes. *Methods Enzymol.* **1987**, *148*, 350–382.

37. García-Limones, C.; Hervas, A.; Navas-Cortes, J.A.; Jimenez-Díaz, R.M.; Tena, M. Induction of an antioxidant enzyme system and other oxidative stress markers associated with compatible and incompatible interactions between chickpea (*Cicer arietinum* L.) and *Fusarium oxysporum* f. sp. *ciceris*. *Physiol. Mol. Plant Pathol.* **2002**, *61*, 325–337. [CrossRef]

38. Yemm, E.M.; Willis, A.J. The estimation of carbohydrates in plant extracts by anthrone. *Biochem. J.* **1954**, *57*, 508–514. [CrossRef]

39. Xu, B.J.; Chang, S.K.C. A comparative study on phenolic profiles and antioxidant activities of legumes as affected by extraction solvents. *J. Food Sci.* **2007**, *72*, S159–S166. [CrossRef]

40. Jia, Z.; Tang, M.C.; Wu, J.M. The determination of flavonoid contents in mulberry and their scavenging effects on superoxide radicals. *Food Chem.* **1999**, *64*, 555–559. [CrossRef]

41. TerBraak, C.F.J.; Šmilauer, P. *CANOCO Reference Manual and User's Guide to Canoco for Windows: Software for Canonical Community Ordination (Version 4)*; Centre for Biometry: Wageningen, The Netherlands, 1998.

42. Chartzoulakis, K. Salinity and olive: Growth, salt tolerance, photosynthesis and yield. *Agric. Water Manag.* **2005**, *78*, 108–121. [CrossRef]

43. Angelopoulos, K.; Dichio, B.; Xiloyannis, C. Inhibition of photosynthesis in olive trees (*Olea europaea* L.) during water stress and rewatering. *J. Exp. Bot.* **1996**, *47*, 1093–1100. [CrossRef]

44. Bongi, G.; Loreto, F. Gas-exchange properties of salt-stressed olive (*Olea europaea* L.) leaves. *Plant Physiol.* **1989**, *90*, 1408–1413. [CrossRef]

45. Evans, G.R.; Von Caemmerer, S.; Setchell, B.A.; Hudson, G.S. The relationship between CO_2 transfer conductance and leaf anatomy in transgenic tobacco with a reduced content of Rubisco. *Austr. J. Plant Physiol.* **1994**, *21*, 475–495. [CrossRef]

46. Syvertsen, J.P.; Lloyd, J.; McConchie, C.; Kriedemann, P.E.; Farquhar, G.D. On the site of biophysical constrains to CO_2 diffusion through the mesophyll of hypostomatous leaves. *Plant Cell Environ.* **1995**, *18*, 149–157. [CrossRef]

47. Tattini, M.; Gucci, R.; Coradeschi, M.A.; Ponzio, C.; Everarard, J.D. Growth, gas exchange and ion content in *Olea europaea* plants during salinity and subsequent relief. *Physiol. Plantarum.* **1995**, *95*, 203–210. [CrossRef]

48. Chartzoulakis, K.; Loupassaki, M.; Bertaki, M.; Androulakis, I. Effects of NaCl salinity on growth, ion content and CO_2 assimilation rate of six olive cultivars. *Sci. Horticult.* **2002**, *96*, 235–247. [CrossRef]

49. Pérez-Alfocea, F.; Albacete, A.; Ghanem, M.E.; Dodd, I.C. Hormonal regulation of source-sink relations to maintain crop productivity under salinity: A case study of root-to-shoot signalling in tomato. *Funct. Plant Biol.* **2010**, *37*, 592–603. [CrossRef]

50. Rejsková, A.; Patková, L.; Stodulková, E.; Lipavská, H. The effect of abiotic stresses on carbohydrate status of olive shoots (*Olea europaea* L.) under in vitro conditions. *J. Plant Physiol.* **2007**, *164*, 174–184. [CrossRef]

51. Smirnoff, N. The role of active oxygen in the response to water deficit and desiccation. *New Phytol.* **1993**, *125*, 27–58. [CrossRef]

52. Dixon, R.A.; Paiva, N.L. Stress-induced phenylpropanoid metabolism. *Plant Cell.* **1995**, *7*, 1085–1097. [CrossRef]

53. Gutierrez-Albanchez, E.; Kirakosyan, A.; Bolling, S.F.; García-Villaraco, A.; Gutierrez-Mañero, J.; Ramos-Solano, B. Biotic elicitation as a tool to improve strawberry and raspberry extract potential on metabolic syndrome-related enzymes in vitro. *J. Sci. Food Agric.* **2019**, *99*, 2939–2946. [CrossRef]

54. Boue, S.M.; Cleveland, T.E.; Carter-Wientjes, C.; Shih, B.Y.; Bhatnagar, D.; Mclachlan, J.M.; Burow, M.E. Phytoalexin-Enriched Functional Foods. *J. Agric. Food Chem.* **2009**, *57*, 2614–2622. [CrossRef]

55. Koudounas, K.; Banilas, G.; Michaelidis, C.; Demoliou, C.; Rigas, S.; Hatzopoulos, P. A defence-related *Olea europaea* β-glucosidase hydrolyses and activates oleuropein into a potent protein cross-linking agent. *J. Exp. Bot.* **2015**, *66*, 2093–2106. [CrossRef]

56. Rivera-Utrilla, J.; Ferro-García, M.A.; Mingorance, M.D.; Bautista-Toledo, I. Adsorption of lead on activated carbons from olive stones. *J. Chem. Technol. Biotechnol.* **1986**, *36*, 47–52. [CrossRef]

57. Galiatsatou, P.; Metaxas, M.; Kasselouri-Rigopoulou, V. Mesoporous activated carbon from agricultural byproducts. *Mikrochim. Acta.* **2001**, *136*, 147–152. [CrossRef]

58. Khalil, L.B.; Girgis, B.S.; Tawfik, T. Decomposition of H_2O_2 on activated carbon obtained from olive stones. *J. Chem. Technol. Biotechnol.* **2001**, *76*, 1132–1140. [CrossRef]

59. Blázquez, G.; Hernainz, F.; Calero, M.; Ruiz-Nuñez, L.F. Removal of cadmium ions with olive stones: The effect of some parameters. *Proc. Biochem.* **2005**, *40*, 2649–2654. [CrossRef]

60. Ortega-Jurado, A.; Palomar-Carnicero, J.M.; Cruz-Peragon, F. Integral olive kernel elaboration with electricity generation compared with actual system of obtaining olive virgin oil. *Grasas y Aceites* **2004**, *55*, 303–311. [CrossRef]

61. Vlyssides, A.G.; Loizides, M.; Karlis, P.K. Integrated strategic approach for reusing olive oil extraction. *J. Clean. Prod.* **2004**, *12*, 603–611. [CrossRef]

 *foods*

Article

New Evidences of Antibacterial Effects of Cranberry Against Periodontal Pathogens

María C. Sánchez [1], Honorato Ribeiro-Vidal [1], Begoña Bartolomé [2], Elena Figuero [1], M. Victoria Moreno-Arribas [2], Mariano Sanz [1] and David Herrera [1,*]

[1] ETEP (Etiology and Therapy of Periodontal Diseases) Research Group, University Complutense of Madrid, Plaza Ramón y Cajal s/n, 28040 Madrid, Spain; mariasb@farm.ucm.es (M.C.S.); hribeiro@ucm.es (H.R.-V.); elfiguer@ucm.es (E.F.); marsan@ucm.es (M.S.)

[2] Institute of Food Science Research (CIAL), CSIC-UAM, c/ Nicolás Cabrera 9, 28049 Madrid, Spain; b.bartolome@csic.es (B.B.); victoria.moreno@csic.es (M.V.M.-A.)

* Correspondence: davidher@odon.ucm.es

Received: 20 January 2020; Accepted: 20 February 2020; Published: 24 February 2020

Abstract: The worrying rise in antibiotic resistances emphasizes the need to seek new approaches for treating and preventing periodontal diseases. The purpose of this study was to evaluate the antibacterial and anti-biofilm activity of cranberry in a validated in vitro biofilm model. After chemical characterization of a selected phenolic-rich cranberry extract, its values for minimum inhibitory concentration and minimum bactericidal concentration were calculated for the six bacteria forming the biofilm (*Streptococcus oralis*, *Actinomyces naeslundii*, *Veillonella parvula*, *Fusobacterium nucleatum*, *Porphyromonas gingivalis*, and *Aggregatibacter actinomycetemcomitans*). Antibacterial activity of the cranberry extract in the formed biofilm was evaluated by assessing the reduction in bacteria viability, using quantitative polymerase chain reaction (qPCR) combined with propidium monoazide (PMA), and by confocal laser scanning microscopy (CLSM), and anti-biofilm activity by studying the inhibition of the incorporation of different bacteria species in biofilms formed in the presence of the cranberry extract, using qPCR and CLSM. In planktonic state, bacteria viability was significantly reduced by cranberry ($p < 0.05$). When growing in biofilms, a significant effect was observed against initial and early colonizers (*S. oralis* ($p \leq 0.017$), *A. naeslundii* ($p = 0.006$) and *V. parvula* ($p = 0.010$)) after 30 or 60 s of exposure, while no significant effects were detected against periodontal pathogens (*F. nucleatum*, *P. gingivalis* or *A. actinomycetemcomitans* ($p > 0.05$)). Conversely, cranberry significantly ($p < 0.001$ in all cases) interfered with the incorporation of five of the six bacteria species during the development of 6 h-biofilms, including *P. gingivalis*, *A. actinomycetemcomitans*, and *F. nucleatum*. It was concluded that cranberry had a moderate antibacterial effect against periodontal pathogens in biofilms, but relevant anti-biofilm properties, by affecting bacteria adhesion in the first 6 h of development of biofilms.

Keywords: polyphenols; cranberry; periodontal diseases; dental biofilm; antibacterial activity; anti-biofilm activity; *F. nucleatum*; *P. gingivalis*; *A. actinomycetemcomitans*

---

1. Introduction

Dental biofilm-organized periodontal pathogens (including *Porphyromonas gingivalis* and *Aggregatibacter actinomycetemcomitans*) are the primary etiological factor of periodontal diseases, which are one of the most prevalent conditions affecting human beings [1]. These conditions have not only a relevant impact in the mouth [1], but also in systemic health [2] and in quality of life indicators [3]. Due to the infectious nature of periodontal diseases, antimicrobials are widely used in their management (prevention and treatment) [4,5]. However, the worrying rise in antibiotic resistances, including those in periodontal pathogens [6] and unwanted effects of antiseptics/antimicrobials compounds [4,5] emphasize the need to seek new approaches for treating and preventing periodontal diseases. Therefore, attention is given to the need of finding, developing, and improving antimicrobial natural compounds, capable of inhibiting the proliferation and/or adhesion of bacteria pathogens in dental/oral biofilms [5,7–10].

In previous studies, it has been shown that polyphenols, and other compounds derived from plants have an influence on human microbiota, either by promoting the growth of beneficial microorganisms or by acting against pathogens [11,12]. Cranberry (*Vaccinium macrocarpum*) compounds, including phenolic acids, proanthocyanidins (particularly, A-type proanthocyanidins), anthocyanins, organic acids, and their microbial-derived metabolites [13], selectively inhibit the growth of intestinal pathogens such as *Staphylococcus strains* and *Salmonella enterica* [14], reduce *Escherichia coli* colonization of the urinary tract [15–17], restrict the virulence of *Pseudomonas aeruginosa* [18,19], present anti-oxidant potential [20], anti-adhesion of Gram-negative and Gram-positive bacteria [21,22], and anti-motility [23,24]. Furthermore, they may be associated with relevant health benefits, including a decreased risk of cardiovascular disease-related mortality [25], prevention of type 2 diabetes mellitus [26], and potential anti-cancer properties [27,28].

The antibacterial and anti-adhesion features of cranberry against oral bacteria have drawn wide attention [22,29–31]. Several in vivo and in vitro studies have evaluated how certain cranberry derived compounds could interfere with formation of a cariogenic biofilm. In this regard, it has been demonstrated that certain components of cranberries may limit dental caries by inhibiting the production of organic acids by cariogenic bacteria, the formation of biofilms by *Streptococcus mutans* and *Streptococcus sobrinus*, and the adhesion and coaggregation of a considerable number of other oral species of *Streptococcus* [32–36]. Focusing on periodontal diseases, the non-dialyzable constituent fraction of cranberry (NDM) inhibits the formation of *P. gingivalis* [37] and *Fusobacterium nucleatum* [38] biofilms, two bacteria species associated with periodontitis. The NDM fraction may also inhibit the adhesion of *P. gingivalis* to various proteins, including type I collagen [37] and may reduce bacterial coaggregation involving periodontal pathogens [32]. However, the information on the antibacterial and anti-biofilm capacity of natural extracts from cranberry against relevant periodontal pathogens, growing in complex multi-species biofilms, is scarce.

Therefore, the aim of the present study was to evaluate the antibacterial and anti-biofilm activity of cranberry extracts in a multispecies in vitro biofilm model, including six bacteria species (*Streptococcus oralis*, *Veillonella parvula*, *Actinomyces naeslundii* and the periodontal pathogens *P. gingivalis*, *A. actinomycetemcomitans*, and *F. nucleatum*). The specific objectives were to assess (1) the antibacterial activity of a cranberry extract against bacteria species in formed biofilms, by assessing the reduction in bacteria viability, and (2) the anti-biofilm activity, by studying the inhibition of the incorporation of different bacteria species in biofilms formed in the presence of the cranberry extract.

2. Materials and Methods

2.1. Cranberry Extract

The cranberry extract used in this study was provided by Triarco Industries Inc. (Cranbury, NJ, USA). For determination of its total polyphenols content, the extract (0.05 g) was dissolved in 10 mL of methanol/HCl (1000:1, *v/v*), sonicated (120 W) for 5 min followed by an extra 15 min resting period, centrifuged, and filtered through a 0.22-μm membrane filter. For analysis of individual phenolic compounds, the extract (0.50 g) was dissolved in 10 mL of MeOH/H_2O (20:80, *v/v*) containing 0.2% HCl, sonicated for 10 min, centrifuged, and filtered through 0.22 μm. In both cases, sample preparations were performed in duplicate.

2.2. Analysis of Phenolic Compounds in the Cranberry Extract

Total polyphenols content was measured by the Folin-Ciocalteu reagent (Merck, Darmstadt, Germany) and using gallic acid (25–500 mg L^{-1}) as a calibration standard. Analysis of individual phenolic compounds was carried out by UPLC-DAD-ESI-TQ MS, as previously described in

Sanchez-Patán et al. [39]. Different phenolic acids (including phenylpropionic, phenylacetic, mandelic, benzoic, and cinnamic acids), flavan-3-ols (monomers, B-type procyanidin dimers and trimers, and A-type procyanidin dimers and trimers), and anthocyanins (peonidin, cyanidin, and malvidin derivatives) were targeted [39]. Commercial standards of these phenolic acids were used to construct calibration curves for sample quantification [39].

2.3. Bacteria Strains and Culture Conditions

Reference strains of *S. oralis* CECT 907T, *V. parvula* NCTC 11810, *A. naeslundii* ATCC 19039, *F. nucleatum* DMSZ 20482, *A. actinomycetemcomitans* DSMZ 8324, and *P. gingivalis* ATCC 33277 were used. These bacteria were grown on blood agar plates (Blood Agar Oxoid No 2; Oxoid, Basingstoke, UK), supplemented with 5% (*v/v*) sterile horse blood (Oxoid), 5.0 mg L^{-1} hemin (Sigma, St. Louis, MO, USA) and 1.0 mg L^{-1} menadione (Merck, Darmstadt, Germany) in anaerobic conditions (10% H$_2$, 10% CO$_2$, and balance N$_2$) at 37 °C for 24–72 h.

2.4. Antibacterial Assays

Figure 1 shows the experimental design followed for the study of the antibacterial effects of cranberry against planktonic bacteria and in an oral biofilm model.

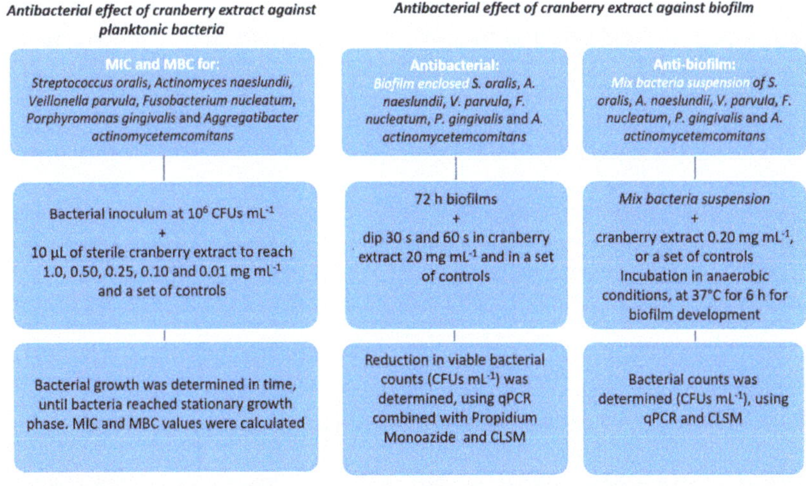

Figure 1. Scheme of the antibacterial assays carried out in this study.

2.4.1. Antibacterial Effect of Cranberry Extract Against Planktonic Bacteria

Pure cultures of the bacteria species were grown anaerobically in a protein rich medium containing brain-heart infusion (BHI) (Becton, Dickinson and Company, Franklin Lakes, NJ, USA) supplemented with 2.5 g L^{-1} mucin (Oxoid), 1.0 g L^{-1} yeast extract (Oxoid), 0.1 g L^{-1} cysteine (Sigma), 2.0 g L^{-1} sodium bicarbonate (Merck), 5.0 mg L^{-1} hemin (Sigma), 1.0 mg L^{-1} menadione (Merck), and 0.25% (*v/v*) glutamic acid (Sigma). The bacteria growth was harvested at mid-exponential phase (measured by spectrophotometry). Microtitre plate-based antibacterial assays were carried out in a 96-wells plate, adding 190 µL of each bacteria inoculum at a final concentration of 10⁶ colony forming units (CFUs) mL^{-1}, and 10 µL of the sterile cranberry extract at a final concentration of 1.0, 0.50, 0.25, 0.10, and 0.01 mg mL^{-1}. Plates had a set of controls: negative control (culture media without any inoculum/cranberry extract), positive control (bacteria without any treatment) as well as blanks (cranberry extract or dimethyl sulfoxide (DMSO) dissolved in the culture media), to ensure the validity of the assay, 4% DMSO (to identify a possible bactericidal effect of DMSO, used as a solvent for the cranberry extract), and 0.2% chlorhexidine (CHX), in order to compare with the reference of known antibacterial effect. A measurement (optical density, O.D.$_{595}$) as t = 0 absorbance was taken in a microtitre plate reader (Optic Ivymen System 2100-C; I.C.T.; La Rioja, Spain). The microplates were incubated for 48 h at 37 °C under anaerobic conditions, and absorbance was measured at selected intervals (1 h during the first 12 h, and every 12 h to complete 48 h), in order to determine the bacteria growth in time, until the bacteria reached stationary growth phase. MIC (minimum inhibitory concentration) and MBC (minimum bactericidal concentration) values were calculated and confirmed by microbial plate counting on blood agar media. Accordingly, the lowest concentration of the cranberry extract showing growth inhibition was considered as the MIC, whereas the lowest concentration of the cranberry extract that showed zero growth in blood agar plates, after spot inoculation and incubation for 72 h, was recorded as the MBC. All experiments were performed in triplicate with appropriate controls.

2.4.2. Antibacterial Effect in an Oral Biofilm Model in Vitro

In order to optimize the method for evaluating the antibacterial effect of the cranberry extract against the bacteria species growing in biofilms, a range of cranberry concentrations were initially tested (from MBCs to stock solution of cranberry extracts at 20 mg mL^{-1}). A dose of 20 mg mL^{-1} yielded the higher antibacterial effect (data not shown).

The multi-species in vitro biofilm model was developed as previously described by Sánchez et al. [40]. Briefly, pure cultures of each bacteria specie were grown anaerobically in the supplemented BHI medium. The bacteria growth was harvested at mid-exponential phase (measured by spectrophotometry), and a mixed bacteria suspension in modified BHI medium containing 10^3 CFUs mL^{-1} for *S. oralis*, 10^5 CFUs mL^{-1} for *V. parvula* and *A. naeslundii*, and 10^6 CFUs mL^{-1} for *F. nucleatum*, *A. actinomycetemcomitans* and *P. gingivalis* was prepared (different concentrations based on the different growth rates of each bacteria species). Sterile calcium hydroxyapatite (HA) discs of 7 mm of diameter and 1.8 mm (standard deviation, SD = 0.2) of thickness (Clarkson Chromatography Products, Williamsport, PA, USA) were coated with sterile saliva for 4 h at 37 °C in sterile plastic tubes to allow the formation of the acquired pellicle [40], and then placed in the wells of a 24-well tissue culture plate (Greiner Bio-one, Frickenhausen, Germany). Each well was inoculated with 1.5 mL mixed bacteria suspension prepared and incubated in anaerobic conditions (10% H$_2$, 10% CO$_2$, and balance N$_2$) at 37 °C for 72 h.

After 72 h, biofilms were dipped during 30 s and 60 s in the cranberry solution (20 mg mL^{-1}), at room temperature. Exposure time of 30 and 60 s were selected since cranberry extracts are bioactive products, commercially available, and for them, the standard exposure times established for other antimicrobial commercially available products (e.g., chlorhexidine mouth rinses), were selected [41–43]. Phosphate buffered saline solution (PBS) was used as negative control and, in order to discard a bactericidal effect of DMSO used to dissolve the extracts, a 4% DMSO solution was also tested.

The antibacterial activity in 72 h biofilms was examined by determining the reduction in the number of viable bacteria counts (expressed as CFUs mL^{-1}), using quantitative polymerase chain reaction (qPCR) combined with Propidium Monoazide (PMA), and by Confocal Laser Scanning Microscopy (CLSM). Assays were conducted in triplicate (with trios of biofilms per replica).

2.5. Anti-Biofilm Assay

In order to optimize the method for evaluating the anti-biofilm effect of cranberry extracts against the selected bacteria species, different concentrations were tested, based on MICs of each bacteria species in planktonic state (data not shown), and it was finally concluded that a dose of 0.20 mg mL^{-1} provided the largest anti-biofilm impact, without affecting bacteria viability in planktonic state.

For the anti-biofilm assay, the mixed bacteria suspension in modified BHI medium containing 10^3 CFUs mL^{-1} for *S. oralis*, 10^5 CFUs mL^{-1} for *V. parvula* and *A. naeslundii*, and 10^6 CFUs mL^{-1} for *F. nucleatum*, *A. actinomycetemcomitans* and *P. gingivalis* was prepared as previously described. HA discs were coated with treated saliva for 4 h at 37 °C in sterile plastic tubes, and then placed in the wells of a 24-well tissue culture plates. Each well was inoculated with 1.5 mL mixed bacteria suspension prepared and the cranberry extract at 0.20 mg mL^{-1}, or with PBS and DMSO in control biofilms, were added. Plates were incubated in anaerobic conditions, at 37 °C for 6 h.

The anti-biofilm activity was examined by determining bacteria counts in biofilms, as CFUs mL^{-1} by means of qPCR, and by CLSM. Assays were conducted in triplicate (with trios of biofilms per replica).

2.6. Microbiological Outcomes

After antibacterial and anti-biofilm assays, biofilms were recovered and sequentially rinsed in 2 mL of sterile PBS (immersion time per rinse, 10 s) three times, in order to remove possible remnants of the extracts and non-adherent bacteria. Then, biofilms were disrupted by vortex for 2 min in 1 mL of PBS. In the case of antibacterial activity, and to discriminate between DNA from live and dead bacteria, PMA was used (Biotium Inc., Hayword, CA, USA). The use of this PMA dye has shown the ability to distinguish between viable and irreversibly damaged cells and hence when combined with qPCR to detect the DNA from viable bacteria [44]. PMA was added to sample tubes containing 250 µL of disaggregated biofilm cells, at a final concentration of 100 µM. Following an incubation period of 10 min at 4 °C in the dark, the samples were subjected to light-exposure for 30 min, using PMA-Lite LED Photolysis Device (Biotium Inc.). After PMA photo-induced DNA cross-linking, the cells were centrifuged at 15,000 rcf for 3 min prior to DNA isolation.

Bacteria DNA was isolated from all biofilms using a commercial kit ATP Genomic DNA Mini Kit® (ATP biotech. Taipei, Taiwan), following manufacturer's instructions and the hydrolysis 5′nuclease probe assay qPCR method was used for detecting and quantifying the bacteria DNA. The qPCR amplification was performed following a protocol previously optimized by our research group, using primers and probes targeted against 16S rRNA gene (obtained through Life Technologies Invitrogen (Carlsbad, CA, USA) and Applied Biosystems (Carlsbad, CA, USA)) [44].

Each DNA sample was analyzed in duplicate. Quantification cycle (Cq) values, previously known as cycle threshold (Ct) values, describing the PCR cycle number at which fluorescence rises above the baseline, were determined using the provided software package (LC 480 Software 1.5; Roche Diagnostic GmbH; Mannheim, Germany). Quantification of viable cells by qPCR was based on standard curves. The correlation between Cq values and CFUs mL^{-1} was automatically generated through the software (LC 480 Software 1.5; Roche).

All assays were developed with a linear quantitative detection range established by the slope range of 3.3–3.7 cycles/log decade, $r^2 > 0.998$, and an efficiency range of 1.9–2.0.

2.7. Confocal Laser Scanning Microscopy (CLSM) Analyses

After antibacterial and anti-biofilm treatment referred above, and before CLSM analysis, the discs were sequentially rinsed in 2 mL of sterile PBS (immersion time per rinse, 10 sec) three times, in order to remove possible remnants of the extract and non-adherent bacteria. Non-invasive confocal imaging of fully hydrated biofilms was carried out using a fixed-stage Ix83 Olympus inverted microscope coupled to an Olympus FV1200 confocal system (Olympus; Shinjuku, Tokyo, Japan). Specimens were stained with LIVE/DEAD® BacLight™ Bacteria Viability Kit solution (Molecular Probes B. V., Leiden, The Netherlands) at room temperature. The 1:1 fluorocrome ratio with a staining time of 9 ± 1 min was used to obtain the optimum fluorescence signal at the corresponding wave lengths (Syto9: 515–530 nm; Propidium Iodide (PI): >600 nm). At least three separate and representative locations on the discs covered with biofilm were selected for these measurements (based on the presence of stacks or "towers" identified in the confocal field view). The CLSM software was set to take a z-series of scans (xyz) of 1 µm thickness (8 bits, 1024 × 1024 pixels). Image stacks were analyzed by using the Olympus[®] software (Olympus). Image analysis and live/dead cell ratio (i.e., the area occupied by living cells divided by the area occupied by dead cells) was performed with Fiji software (ImageJ Version 2.0.0-rc-65/1.52b, Open source image processing software).

2.8. Statistical Analyses

The selected outcome variables to study the antibacterial effect of cranberry extracts were the counts of viable bacteria present on the biofilms, expressed as viable CFUs mL^{-1} of *S. oralis*, *V. parvula*, *A. naeslundii*, *F. nucleatum*, *A. actinomycetemcomitans*, and *P. gingivalis* by qPCR, and the live/dead cell ratio of the whole biofilm by CLSM. An experiment-level analysis was performed for each parameter of the study (n = 9 for qPCR and n = 3 for CLSM results). Shapiro–Wilk goodness-of-fit tests and distribution of data were used to assess normality. The effect of each solution (cranberry extracts, PBS and 4% DMSO), the time of exposure (30 or 60 s), and their interaction with the main outcome variable (counts expressed as CFUs mL^{-1} or live/dead cell ratio), was compared by means of a parametric ANOVA test for independent samples, and a general linear model was constructed for each bacterium for qPCR results and for total bacteria for live/dead cell ratio of whole biofilm obtained by CLSM, using the method of maximum likelihood and Bonferroni corrections for multiple comparisons.

To study the anti-biofilm effect of the cranberry extract, the selected outcome variables were the counts of bacteria present on the biofilms, expressed as CFUs mL^{-1} of *S. oralis*, *V. parvula*, *A. naeslundii*, *F. nucleatum*, *A. actinomycetemcomitans*, and *P. gingivalis* by qPCR, and the live/dead cell ratio of the whole biofilm by CLSM. Shapiro–Wilk goodness-of-fit tests and distribution of data were used to assess normality. An experiment-level analysis was performed for each parameter of the study (n = 9 for qPCR and n = 3 for CLSM results). The effect of each solution (cranberry extract, PBS, and 4% DMSO) on the main outcome variables (CFUs mL^{-1} or live/dead cell ratio), was compared by means of a parametric ANOVA test for independent samples, using the method of maximum likelihood and Bonferroni corrections for multiple comparisons.

Data was expressed as means ± SD and as the mean percent inhibition that was calculated by Equation: Percent inhibition = (CFUs mL^{-1} of negative control–CFUs mL^{-1} of test / CFUs mL^{-1} of negative control) × 100.

Results were considered statistically significant at $p < 0.05$. A software package (IBM SPSS Statistics 24.0; IBM Corporation, Armonk, NY, USA) was used for all data analysis.

3. Results

3.1. Phenolic Composition of the Cranberry Extract

A phenolic characterization of the cranberry extract was initially carried out to ensure its susceptibility for this study. The content in total polyphenols content of the extract resulted in 219 mg of gallic acid equivalents g^{-1}. Concerning phenolic composition, Table 1 details the main phenolic compounds found in the extract, as determined by UPLC-DAD-ESI-TQ MS. Among phenolic acids (benzoic and cinnamic acids), the extract was especially rich in benzoic acid (8.38 mg g^{-1}), followed by others such as p-coumaric acid (0.84 mg g^{-1}) and protocatechuic acid (0.73 mg g^{-1}) in considerable less content. Concerning flavan-3-ols, main MS signals corresponded to A-type trimers (1.58 mg g^{-1}), followed by A-type (0.23 mg g^{-1}) and B-type (0.20 mg g^{-1}) dimers, monomers (0.065 mg g^{-1}), and B-type trimers (0.034 mg g^{-1}). In relation to anthocyanins, peonidin-3-arabinoside (0.32 mg g^{-1}) and cyanidin-3-arabinoside (0.15 mg g^{-1}) showed the highest content (Table 1). These compositional data were in accordance to others commercial cranberry extracts [39].

Table 1. Phenolic compounds present in the cranberry extract used in this study. Data are expressed as mean and standard deviation (SD).

Compounds Group	Phenolic Compound	Concentration (μg g^{-1} ± SD)
Benzoic acids	Benzoic acid	8317.88 ± 222.31
	Protocatechuic acid	735.12 ± 17.76
	Vanillic acid	262.54 ± 10.16
	Gallic acid	136.16 ± 1.50
	4-Hydroxybenzoic acid	94.81 ± 2.23
	Salycilic acid	91.05 ± 2.16
	4-Hydroxymandelic acid	30.84 ± 1.14
	3-O-methylgallic acid	30.05 ± 0.64
	4-Hydroxy-3-methoxymandelic acid	14.33± 0.45
	Syringic acid	11.80 ± 1.18
	3-Hydroxybenzoic acid	11.58 ± 0.01
	3-(3,4-Dihydroxyphenyl)-propionic acid	9.61 ± 0.16
	4-Hydroxy-3-methoxyphenylacetic acid	6.55 ± 0.66
	3,4-Dihydroxy mandelic acid	3.56 ± 0.31
	4-Hydroxypheny lacetic acid	3.20 ± 0.41
	Hippuric acid	1.14 ± 0.10
	3,4-Dihydroxy phenylacetic acid	1.05 ± 0.10
	3,4,5-Trimethoxy benzoic acid	0.32 ± 0.03
Cinnamic acids	*p*-Coumaric acid	844.16 ± 15.20
	trans-Cinnamic acid	260.55 ± 0.04
	Caffeic acid	133.67 ± 2.52
	Ferulic acid	111.92 ± 4.38
	Trimethoxycinnamic acid	2.72 ± 0.27
Flavan-3-ols	Σ A-type trimers	1579.04 ± 27.31
	Σ A-type dimers	230.95 ± 18.11
	Σ B-type dimers	201.87 ± 17.21
	Σ Monomers	65.81 ± 5.20
	Σ B-type trimers	34.1 ± 0.91
Anthocyanins	Peonidin-3-arabinoside	32.73 ± 3.27
	Cyanidin-3-arabinoside	15.01 ± 0.05
	Peonidin-3-glucoside	4.84 ± 0.48
	Malvidin-3-arabinoside	1.16 ± 0.02
	Peonidin-3-galactoside	1.03 ± 0.09
	Cyanidin-3-glucoside	0.31 ± 0.02
	Cyanidin-3-galactoside	0.19 ± 0.01

3.2. Antibacterial Assays

3.2.1. Antibacterial Effect of Cranberry Extract Against Planktonic Bacteria

MICs and MBCs values against the six bacteria species selected in planktonic state were determined for the selected cranberry extract. MICs indicated an average bacteriostatic concentration of 0.10 mg mL^{-1} against *P. gingivalis* and *F. nucleatum*, 0.25 mg mL^{-1} for *A. naeslundii* and *A. actinomycetemcomitans*, 0.50 mg mL^{-1} for *V. parvula*, and >1.00 mg mL^{-1} for *S. oralis*. MBCs tests showed similar results, with bactericidal concentrations of 0.25 mg mL^{-1} against *P. gingivalis*, 1.00 mg mL^{-1} against *F. nucleatum*, and >1.00 mg mL^{-1} for *S. oralis*, *A. naeslundii*, *V. parvula*, and *A. actinomycetemcomitans*. According to these results, the cranberry extract exhibited antibacterial activity, displaying the largest antibacterial properties against the periodontal pathogens *P. gingivalis* and *F. nucleatum*.

3.2.2. Antibacterial Effects in an in Vitro Biofilm Model: Bacteria Counts

Table 2 depicts the effect of cranberry extracts (20 mg mL^{-1}), compared to the negative control (PBS) and 4% DMSO control solution, on the mean number of viable bacteria counts in 72 h biofilms.

After an exposure of 30 or 60 s to the cranberry extract, significant reductions in viable counts in biofilms were observed for initial and early colonizers. *S. oralis* showed significant reductions after 30 ($p < 0.001$) and 60 s ($p = 0.017$) when compared to negative control (PBS), reaching in both cases a decrease of 98.9% of viable CFUs (Table 2). Significant differences ($p < 0.001$ after 30 s of exposure) were also observed when the effects of DMSO solution was compared to PBS, with percentages of decrease of 93.1% and 58.8% for 30 and 60 s exposures, respectively (Table 2). For *A. naeslundii* and *V. parvula*, a significant impact of the cranberry extract was observed after 30 s (65.7% of reduction, $p = 0.006$ and 66.7% of reduction, $p = 0.010$, respectively), but not after 60 s. No significant reductions were observed after exposure to DMSO ($p > 0.05$), after 30 or 60 s (Table 2). No statistically significant differences were observed between the cranberry extract and DMSO at any time ($p > 0.05$). The effect of exposure time (30 s versus 60 s) was not statistically significant for both solutions ($p > 0.05$ in all cases) in *S. oralis*, *A. naeslundii* and *V. parvula*.

For the secondary colonizer *F. nucleatum*, some effects on viable counts were observed after 30 s ($p = 0.164$) and after 60 s (decrease of 75.3%, $p = 0.448$), although not statistically significant. Additionally, no statistically significant reductions in viable counts were observed for DMSO ($p > 0.05$) (Table 2). No statistically significant differences were observed between the cranberry extract and DMSO at any time ($p > 0.05$). The effect of exposure time was however, statistically significant for the cranberry extract ($p = 0.022$) and DMSO ($p = 0.035$).

For the periodontal pathogens *A. actinomycetemcomitans* and *P. gingivalis*, no significant reductions in viable counts after 30 s or 60 s of exposure to the cranberry extract ($p > 0.05$) were observed when compared to negative control: reductions of 11.5% for *A. actinomycetemcomitans* and 39.3% for *P. gingivalis* after 60 s. The same was true for DMSO ($p > 0.05$) (Table 2). No statistically significant differences were observed in the effectiveness comparing the two solutions at applied times or when comparing exposure times ($p > 0.05$ for all cases).

3.2.3. Antibacterial Effects in an in Vitro Biofilm Model: CLSM

The CLSM analysis showed that, after 72 h of incubation on HA surfaces, control biofilms covered the entire disc surface as a flat layer of cells combined with stacks of bacteria aggregations, showed a live/dead cell ratio (i.e., the area occupied by living cells divided by the area occupied by dead cells) of 1.43 (SD 0.10) and 1.25 (SD 0.15), after exposure of 30 and 60 s, respectively, to PBS (Figure 2a, b). Table 3 depicts the effects of the cranberry extract on the live/dead cell ratio of the whole biofilm obtained by CLSM. It could be observed that, after exposure of 30 s to cranberry extracts and to the 4% DMSO solution, cell vitality significantly decreased in the biofilms, showing, respectively, live/dead cell ratios of 0.67 (SD 0.07) and 0.77 (SD 0.04) for 4% DMSO ($p < 0.001$ in both cases, when compared to negative control biofilms) (Figure 2c, e). After 60 s of exposure (Figure 2f), reductions in viability were also statistically significant for cranberry extracts (live/dead cell ratio of 0.56 (SD 0.02), $p < 0.001$; Figure 2f) and for DMSO solution (live/dead cell ratio of 0.78 (SD 0.05), $p < 0.001$; Figure 2d), when compared to control biofilms (live/dead cell ratio of 1.25 (SD 0.15)). These results are consistent with those observed by means of qPCR, with significant differences in viable counts of initial and early colonizers, after exposure to cranberry extracts and DMSO solution, when compared to negative control biofilms. Statistically significant differences were observed between the cranberry extract and DMSO after 60 s of exposure ($p = 0.027$) (Table 3).

Table 2. Antibacterial effects of the cranberry extract on the mean number of viable bacteria counts in the in vitro multi-species biofilm model (in colony forming units, CFUs mL^{-1}, determined by quantitative polymerase chain reaction (qPCR)). Data are expressed as mean and standard deviation (SD). PBS: phosphate buffer saline; DMSO: 4% dimethyl sulfoxide solution.

	Exposure Time (seconds)	Viable CFUs mL^{-1} [mean (SD)]			p-Value When Compared to Negative Control		% of Reduction of viable CFUs mL^{-1} as Compared with Negative Control	
		Negative Control (PBS)	Cranberry Extract	DMSO	Cranberry Extract	DMSO	Cranberry Extract	DMSO
S. oralis	30	$1.2 \times 10^6 \pm 1.1 \times 10^6$	$1.3 \times 10^4 \pm 1.1 \times 10^4$	$8.3 \times 10^4 \pm 1.4 \times 10^5$	0.000	0.000	98.9	93.1
	60	$6.8 \times 10^5 \pm 4.3 \times 10^5$	$7.3 \times 10^3 \pm 4.4 \times 10^3$	$2.8 \times 10^5 \pm 2.3 \times 10^5$	0.017	0.282	98.9	58.8
A. naeslundii	30	$6.7 \times 10^4 \pm 5.6 \times 10^4$	$2.3 \times 10^4 \pm 1.3 \times 10^4$	$3.4 \times 10^4 \pm 2.1 \times 10^4$	0.006	0.050	65.7	49.2
	60	$2.2 \times 10^4 \pm 1.3 \times 10^4$	$3.2 \times 10^4 \pm 2.4 \times 10^4$	$2.0 \times 10^4 \pm 1.4 \times 10^4$	1.000	1.000	45.4	9.1
V. parvula	30	$3.6 \times 10^6 \pm 2.8 \times 10^6$	$1.2 \times 10^6 \pm 1.4 \times 10^6$	$2.0 \times 10^6 \pm 2.1 \times 10^6$	0.010	0.147	66.7	44.4
	60	$1.6 \times 10^6 \pm 8.6 \times 10^5$	$4.4 \times 10^5 \pm 3.6 \times 10^5$	$1.3 \times 10^6 \pm 1.3 \times 10^6$	0.395	1.000	72.5	18.7
A. actinomycetemcomitans	30	$7.2 \times 10^6 \pm 6.4 \times 10^6$	$6.8 \times 10^6 \pm 4.7 \times 10^6$	$5.6 \times 10^6 \pm 3.0 \times 10^6$	1.000	1.000	5.6	22.2
	60	$5.2 \times 10^6 \pm 3.5 \times 10^6$	$4.6 \times 10^6 \pm 4.4 \times 10^6$	$5.2 \times 10^6 \pm 4.9 \times 10^6$	1.000	1.000	11.5	0.0
P. gingivalis	30	$1.7 \times 10^6 \pm 7.0 \times 10^5$	$1.1 \times 10^6 \pm 5.2 \times 10^5$	$1.6 \times 10^6 \pm 1.8 \times 10^6$	0.434	1.000	35.3	5.9
	60	$8.9 \times 10^5 \pm 6.8 \times 10^5$	$5.4 \times 10^5 \pm 1.8 \times 10^5$	$1.0 \times 10^6 \pm 7.0 \times 10^5$	1.000	1.000	39.3	12.3
F. nucleatum	30	$3.8 \times 10^5 \pm 3.1 \times 10^5$	$2.3 \times 10^5 \pm 1.5 \times 10^5$ †	$3.5 \times 10^5 \pm 1.3 \times 10^5$	0.164	1.000	39.5	7.9
	60	$1.5 \times 10^5 \pm 1.0 \times 10^5$	$3.7 \times 10^4 \pm 3.0 \times 10^4$ †	$1.8 \times 10^5 \pm 1.5 \times 10^5$	0.448	1.000	75.3	0.0

† $p < 0.05$, significant differences when comparing exposure times for an antimicrobial agent.

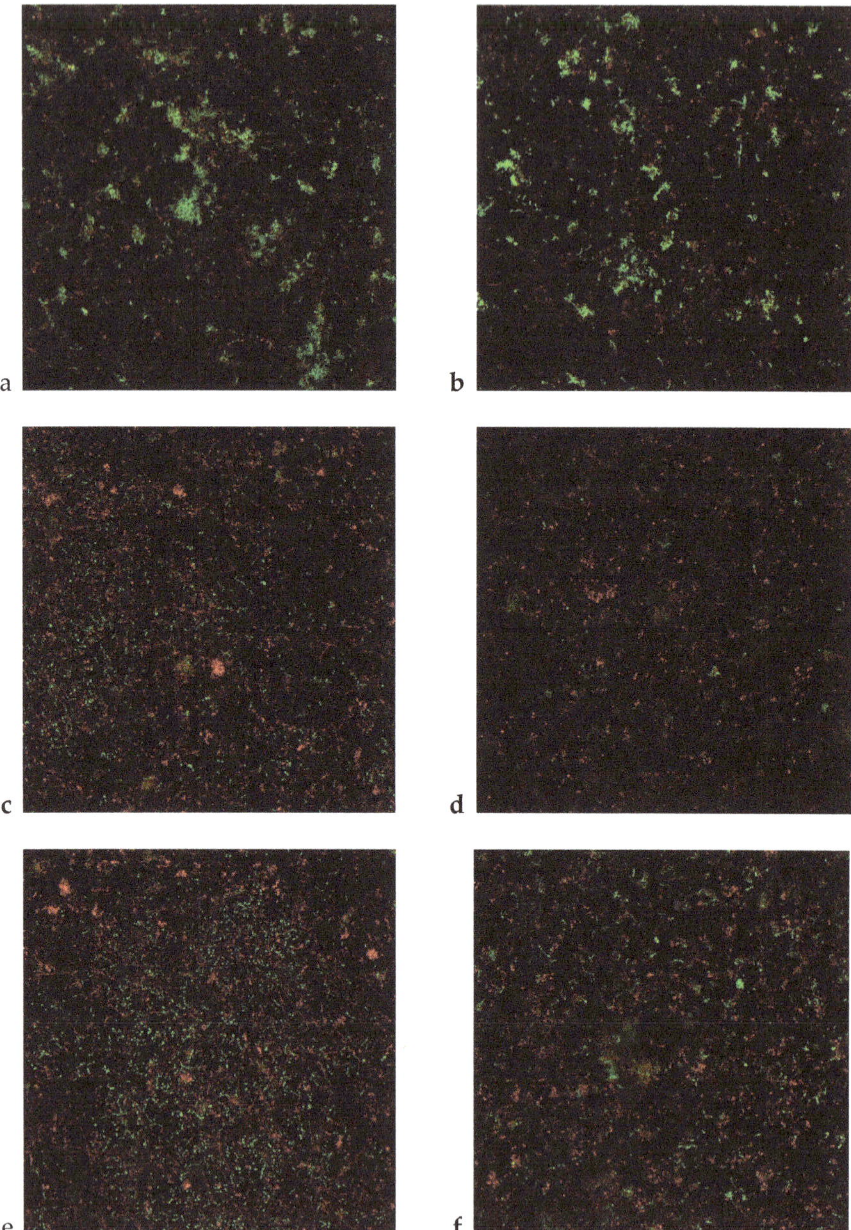

Figure 2. Maximum projection of confocal laser scanning microscopy (CLSM) images of the whole biofilm, grown 72 h over hydroxyapatite surfaces, and stained with LIVE/DEAD® BacLight™ Bacteria Viability Kit, after exposure to: (**a**,**b**) negative controls, 30 and 60 s, respectively (phosphate buffer saline, PBS); (**c**,**d**) 4% dimethyl sulfoxide (DMSO) solution, 30 and 60 s, respectively; (**e**,**f**) cranberry extracts (20 g L^{-1}), 30 and 60 s, respectively. (Scale bar = 100 μm).

Table 3. Effect of the cranberry extract on the live/dead cell ratio (i.e., the area occupied by living cells divided by the area occupied by dead cells) of the whole biofilm obtained by confocal laser scanning microscopy (CLSM). PBS: phosphate buffer saline; DMSO: 4% dimethyl sulfoxide solution.

Treatment			Mean Difference (I–J)	Standard Error	Sig.[a]	95% Confidence Interval for Difference	
						Lower Bound	Upper Bound
Antimicrobial effect							
30 s	PBS	Cranberry	0.763	0.071	0.000	0.567	0.960
		DMSO	0.663	0.071	0.000	0.467	0.860
	Cranberry	DMSO	−0.100	0.071	0.550	−0.297	0.097
60 s	PBS	Cranberry	0.687	0.071	0.000	0.490	0.883
		DMSO	0.467	0.071	0.000	0.270	0.663
	Cranberry	DMSO	−0.220	0.071	0.027	−0.417	−0.023
Anti-biofilm effect							
6 h	PBS	Cranberry	0.35000	0.14575	0.160	−0.1292	0.8292
		DMSO	0.40000	0.14575	0.101	−0.0792	0.8792
	Cranberry	DMSO	0.05000	0.14575	1.000	−0.4292	0.5292

Based on estimated marginal means; [a] *p* value, adjustment for multiple comparisons (Bonferroni).

3.3. Anti-Biofilm Assay

3.3.1. Anti-Biofilm Assay: Bacteria Counts

The cranberry extract, at a concentration of 0.20 mg mL^{-1}, significantly inhibited the incorporation of five of the six studied bacteria species in the in vitro biofilm model (Table 4). After 6 h of contact, and compared to negative control biofilms, two of the three initial and early colonizers were significantly reduced on the HA surfaces: 98.9% for *S. oralis* ($p < 0.001$) or 90.9% for *V. parvula* ($p < 0.001$), when exposed to cranberry extracts. No significant impact was observed for *A. naeslundii*.

Periodontal pathogens showed a similar trend. *P. gingivalis* showed the largest impact of cranberry extracts: 97.2% ($p < 0.001$), with counts of 1.1×10^3 (SD 1.1×10^3) CFUs mL^{-1}, compared to 4.0×10^4 (SD 2.9×10^4) CFUs mL^{-1}, in negative control biofilms. Reductions *A. actinomycetemcomitans* (84.0%) and *F. nucleatum* (75.4%) were statistically significant ($p < 0.001$ in both cases).

For DMSO, a significant impact was observed for the three periodontal pathogens and for *S. oralis*, when compared to control biofilms ($p < 0.005$ in all cases; Table 4).

Significant differences were observed in the effectiveness comparing the cranberry extract and DMSO solution after 6 h of biofilm evolution in *V. parvula* ($p < 0.001$) and *A. actinomycetemcomitans* ($p = 0.024$).

Table 4. Anti-biofilm effects of the cranberry extract on the mean number of bacteria counts, incorporated during the 6 h of devolvement in the in vitro multi-species biofilm model (in colony forming units, CFUs mL^{-1}, determined by quantitative real-time polymerase chain reaction (qPCR)). Data are expressed as mean and standard deviation (SD). PBS: phosphate buffer saline; DMSO: 4% dimethyl sulfoxide solution.

	Viable CFUs mL^{-1} [mean (SD)]			p-Value When Compared to Negative Control		% of Reduction of Viable CFUs mL^{-1} Respect to Negative Control	
	Negative Control (PBS)	Cranberry Extract	DMSO	Cranberry Extract	DMSO	Cranberry Extract	DMSO
S. oralis	$1.2 \times 10^5 \pm 2.5 \times 10^4$	$1.3 \times 10^3 \pm 5.3 \times 10^2$	$5.5 \times 10^2 \pm 2.6 \times 10^2$	0.000	0.000	98.9	99.5
A. naeslundii	$4.8 \times 10^4 \pm 3.1 \times 10^4$	$7.8 \times 10^4 \pm 7.6 \times 10^4$	$6.4 \times 10^4 \pm 1.9 \times 10^4$	0.608	1.000	-	-
V. parvula	$2.3 \times 10^4 \pm 1.5 \times 10^4$	$2.1 \times 10^3 \pm 2.2 \times 10^3$	$2.0 \times 10^4 \pm 7.3 \times 10^3$	0.000	1.000	90.9	13.0
A. actinomycetemcomitans	$7.5 \times 10^5 \pm 2.8 \times 10^5$	$1.2 \times 10^5 \pm 9.5 \times 10^4$	$3.8 \times 10^5 \pm 1.4 \times 10^5$	0.000	0.001	84.0	50.7
P. gingivalis	$4.0 \times 10^4 \pm 2.9 \times 10^4$	$1.1 \times 10^3 \pm 1.1 \times 10^3$	$1.0 \times 10^4 \pm 9.9 \times 10^3$	0.000	0.0047	97.2	75.0
F. nucleatum	$1.1 \times 10^5 \pm 3.8 \times 10^4$	$2.7 \times 10^4 \pm 2.0 \times 10^4$	$5.9 \times 10^4 \pm 2.0 \times 10^4$	0.000	0.005	75.4	46.4

3.3.2. Anti-Biofilm Assay: CLSM

CLSM analysis showed that, after 6 h of incubation on HA surfaces, formed biofilms showed the typical features of bacteria communities in their first steps, with a high percentage of live cells versus dead cells, that was evidenced by a live/dead cell ratio of 1.44 (SD 0.01) (Figure 3a,b). The effect of the exposure of the biofilms for 6 h to the cranberry extract, at a concentration of 0.20 mg mL^{-1}, was evident as it was not possible to observe well-structured biofilms, contrary what happened in the control samples. Although the biomass was reduced, no significant differences in bacteria vitality were observed when compared respect to controls (live/dead cell ratio of 0.99 (SD 0.01), $p = 0.160$; Table 3; Figure 3c,d), suggesting a limited antiseptic effect, and highlighting a desired effect on bacteria adhesion. Conversely, DMSO showed a similar live/dead cell ratio (1.047 (SD 0.14); Figure 3e,f), and biofilms were normally formed, with no significant differences when compared to control biofilms ($p = 0.101$; Table 3). These results are consistent with those observed by qPCR, which showed significant differences in bacteria counts of the tested bacteria species when biofilms formed in the presence of the cranberry extract where compared with control biofilms.

No statistically significant differences were observed in the effectiveness comparing the cranberry extract and DMSO at applied time ($p = 1.000$) (Table 3).

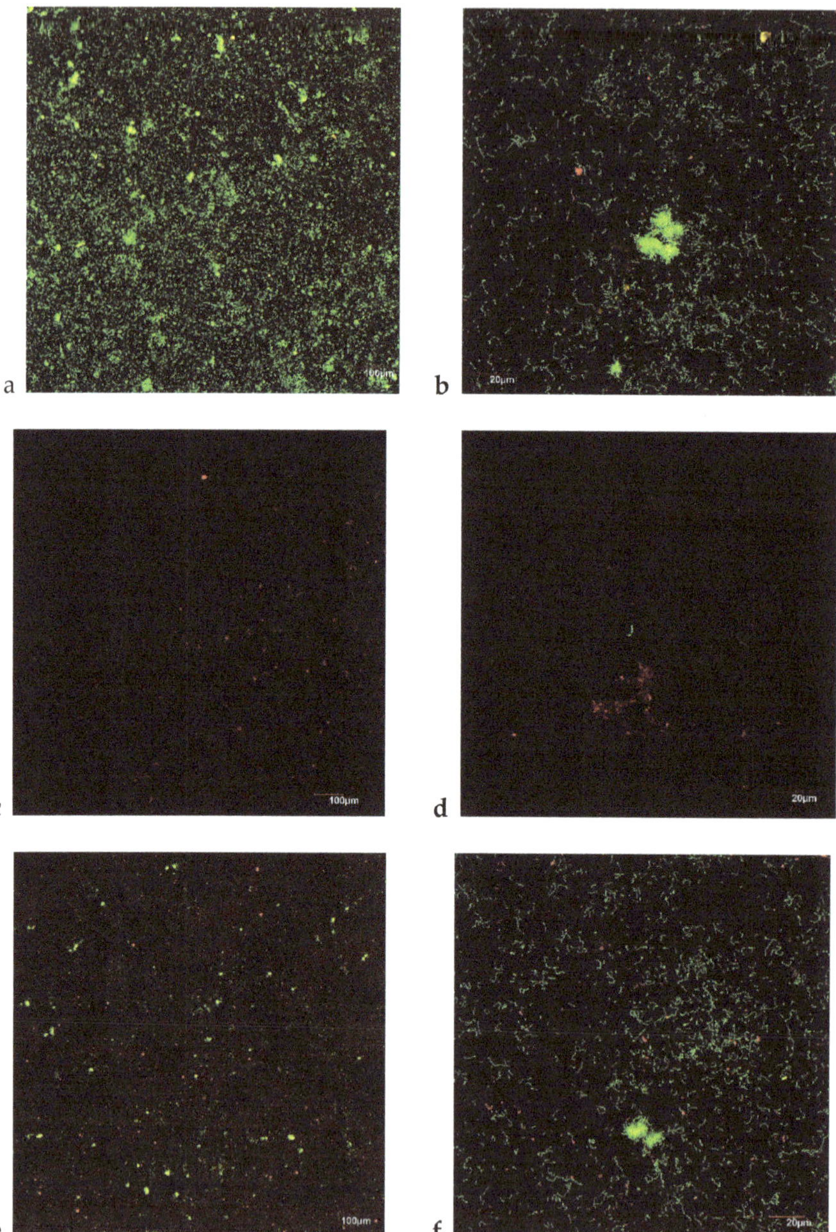

Figure 3. Maximum projection of confocal laser scanning microscopy (CLSM) images of the whole biofilm after 6 h of development, growing in the presence of 0.20 mg mL^{-1} of cranberry extract, over hydroxyapatite surfaces, and stained with LIVE/DEAD® BacLight^{TM} Bacteria Viability Kit, after exposure to: (**a**,**b**) negative control (phosphate buffer saline, PBS); (**c**,**d**) cranberry extract; (**e**,**f**) 4% dimethyl sulfoxide (DMSO) solution.

4. Discussion

Since bacteria resistance to antibiotics is becoming an increasing health threat worldwide, alternative strategies to prevent or limit biofilm formation are a relevant goal. A growing body of evidence has demonstrated that plant extracts offer relevant antimicrobial and anti-biofilm potentials, with no significant risk of increasing antibiotic resistances. A vast number of phytochemicals have been recognized as valuable alternatives and complementary products to manage bacterial infections [45,46]. Cranberry (*Vaccinium macrocarpum*) fruits are particularly rich in biologically active phenolic compounds and organic acids [13], as it has also been confirmed from our results (Table 1). Numerous in vivo and in vitro studies have showed that different cranberry compounds/fractions/extracts possess antibacterial properties (against both Gram-positive and Gram-negative bacteria species) on various pathogenic bacteria in urinary tract infections and other diseases [33,47–49]. In this context, the present study has confirmed the antibacterial capacity of cranberry extracts against the six bacteria species (*S. oralis* CECT 907T, *V. parvula* NCTC 11810, *A. naeslundii* ATCC 19039, *F. nucleatum* DMSZ 20482, *A. actinomycetemcomitans* DSMZ 8324, and *P. gingivalis* ATCC 33277) tested in planktonic state; this findings contradict, at least partially, those of La and co-workers [50], who concluded that A-type proanthocyanidins did not present any effect on *P. gingivalis* in planktonic state. In view of our results from the UPLC-DAD-ESI-TQ MS analysis of the cranberry extract, a possible explanation for this disagreement could be that the antiseptic effect comes, not only from these A-type proanthocyanidins, but also from some of the other components of cranberry extracts, such as phenolic acids. In this context, the necessity of carrying out previous compositional characterization of cranberry extracts should be noted—as we have done in our study—in view of the diversity in this kind of products that affects their bioactivity [51].

However, bacteria are normally arrange as biofilms, predominating these sessile communities in most of the environmental, industrial, and medical habitats [52]. In fact, these highly structured bacteria communities are found in the mouth, allowing bacteria cells to withstand the natural defence mechanisms, as well as the host's immune defences or the effects of antimicrobial agents [53–55]. Therefore, the study of the antimicrobial of cranberry extracts should be performed against bacteria organized in biofilms. The results of the present study indicate that, when testing bacteria organized in biofilms, bacteria viability was affected by exposure to the cranberry extract at 20 mg mL^{-1} after 30 and 60 s of exposure. However, a significant effect was only observed for initial and early colonizers (*S. oralis*, *A. naeslundii*, and *V. parvula*), but, in agreement with other studies, not for periodontal pathogens (*F. nucleatum*, *P. gingivalis*, and *A. actinomycetemcomitans*). Philips and coworkers [56], in a recent investigation assessing the inhibitory effects of berry fruit extracts on *S. mutans* biofilms, indicated that bacteria viability was not significantly affected, as also concluded by Koo and coworkers [33]. Biofilms are an intriguing structure which demonstrate greater resistance to antimicrobial agents when compared to organisms in planktonic form [31]. A previous study using a cranberry juice concentrate formulated as a thermoreversible gel [11], showed antibacterial properties against *A. actinomycetemcomitans* and *P. gingivalis*, in contrast to the results of our study. The variability of the results may be due to the different types of samples and formulations used.

Besides the antibacterial effects, this investigation highlights new possible features regarding the anti-biofilm activity of cranberry extracts against periodontal pathogens. Bacteria adhesion to oral surfaces is the initial and crucial step in dental biofilm development and, therefore, in the pathogenesis of periodontal diseases. The cranberry extract, at a concentration of 0.20 mg mL^{-1}, inhibited the colonization of the six tested bacteria species in the in vitro biofilm model, especially for periodontal pathogens *P. gingivalis* (97.2% of reduction), *A. actinomycetemcomitans* (84%), and *F. nucleatum* (75.4%), being the impact statistically significant ($p < 0.001$ in all cases), when compared to control biofilms. Additionally, initial and early colonizers were significantly affected: *S. oralis* (98.9%, $p < 0.001$) or *V. parvula* (90.9%, $p < 0.001$). Different studies have described the role of cranberry constituents in bacteria adhesion and biofilm development: Philips and coworkers [56] indicated that cranberry extracts were the most effective extract in disrupting *S. mutans* biofilm integrity and structural architecture, without significantly affecting bacteria viability; La and co-workers [50] observed that A-type cranberry proanthocyanidins did not have any effect on *P. gingivalis* planktonic growth, but they did inhibit biofilm formation. The anti-biofilm effect of cranberry extracts in our biofilm model was also confirmed by CLSM, with a significant disturbance on biofilm structure, a qualitative assessment that was consistent with the quantitative data provided by qPCR.

Labreque and coworkers [37] and Yamanaka and coworkers [38] observed that the non-dialyzable constituent fraction of cranberry (NDM) interfered with the colonization of *P. gingivalis* and *F. nucleatum* in the gingival crevice, reducing bacteria coaggregation in periodontal diseases [37,38,57,58]. Moreover, Polak et al. [58] found that NDM adhesion of *P. gingivalis* and *F. nucleatum* onto epithelial cells, and NDM consumption by mice attenuated the severity of experimental periodontitis, compared with a mixed infection without NDM treatment. Furthermore, NDM increased the phagocytosis of *P. gingivalis*. In addition, cranberries were described to restrain the proteolytic activity of the red complex, specifically the gingipain activity of *P. gingivalis*, trypsin-like activity of *Tannerella forsythia*, and chemotrypsin-like activity of *Treponema denticola* [59]. Cranberry extracts have also demonstrated the inhibition of the productions some cytokines: Bodet et al. [59] or Polak et al. [58] observed that NDM eliminated TNF-a expression by macrophages that were exposed to *P. gingivalis* and *F. nucleatum*, without impairing their viability.

The hydrophobic character of the cranberry extract has made the experiments difficult, requiring the use of the organic solvent DMSO in the tests, in order to overcome such complications. However, some antimicrobial activity of DMSO at the selected concentration (4%) was observed, and therefore, it may have a possible contribution in the antibacterial activity of the extract under investigation. In this way, studies have tested different concentrations, ranging up to 10% [60–63]. However, when used as a solvent, there is no established criteria as to which is the most appropriate concentration, and the interpretation of its effects on the microorganisms with which it interacts is of great importance in view of its widespread use as solvent in therapeutic and pharmacological studies [60–63]. In the present study, the 4% DMSO concentration was selected as the one that ensured complete solubilisation of the cranberry extract with minimum antimicrobial effects. However, in any case, the results obtained in the present study make evident the need to standardize an appropriate concentration of DMSO, suitable for bacterial experiments, considering that there is a discrepancy in the findings of different studies on the antimicrobial effects of different concentrations of DMSO.

5. Conclusions

This study has demonstrated that the incorporation of bacteria into the biofilm was significantly interfered, including relevant periodontal pathogens, such as *P. gingivalis*, *A. actinomycetemcomitans* and *F. nucleatum*. Our results support the hypothesis that cranberry components may interfere in the phase of bacteria adherence, disabling or inhibiting the adherence of periodontal pathogens and, therefore, preventing bacterial colonization. This fact could interfere with biofilm formation and possibly helping to maintain homeostasis and, thus, to prevent periodontal diseases. Anti-biofilm activity of cranberry extracts in the present study could be attributed to the presence of polyphenols, specifically phenolic acids and A-type proanthocyanidins, which are known to inactivate glucosyl-transferase and fructosyl-transferase that catalyse the formation of glucan and fructan, respectively, which play prime roles in biofilm formation and maturation [31]. It has also been reported that the polyphenols in cranberries led to desorption of biofilm by interfering with bacteria coaggregation [64]. Moreover, cranberries are supposed to reduce periodontal-related symptoms by suppressing inflammatory cascades as an immunologic response to bacteria invasion.

Despite the limitations of the study, and the great effect caused by the DMSO solvent, the research performed has identified an important anti-biofilm effect of cranberry on periodontal bacteria and serve as a support for the development of further studies, assessing the most effective vehicle and the ideal concentration to be used, without causing adverse effects on oral tissues.

Author Contributions: M.C.S. and H.R.-V. contributed to conception and design of the study with the aid of E.F., B.B., M.V.M.-A., M.S., and D.H., analysis and interpretation of data and drafted the manuscript. E.F. performed the statistical analyses. E.F., B.B., M.V.M.-A., M.S., and D.H. critically revised the manuscript. All authors reviewed the original draft and read and approved the final manuscript.

Funding: This work was funded by the Spanish Ministry of Industry, Economy and Competitiveness (AGL2015–64522-C2-R project) and the Comunidad de Madrid (ALIBIRD-CM 2020 P2018/BAA-4343). It was also part of the activities of Dentaid Extraordinary Chair in Periodontal Research (*Cátedra Extraordinaria Dentaid en Investigación Periodontal*, University Complutense of Madrid, Spain).

Acknowledgments: We thank for their technical assistance to Luis M. Alonso and Alfonso Cortés, from the Centre of Microscopy and Cytometry from the Complutense University of Madrid.

Conflicts of Interest: The authors declare no conflict of interest. None of the funders had any role in designing and/or conducting of the study; collection, management, analysis and interpretation of the data; and preparation, review, or approval of the manuscript.

References

1. Teles, R.; Teles, F.; Frias-Lopez, J.; Paster, B.; Haffajee, A. Lessons learned and unlearned in periodontal microbiology. *Periodontol* **2013**, *62*, 95–162. [CrossRef]
2. Tonetti, M.S.; Van Dyke, T.E. Working group 1 of the joint EFPAAPw. Periodontitis and atherosclerotic cardiovascular disease: Consensus report of the Joint EFP/AAP Workshop on Periodontitis and Systemic Diseases. *J. Periodontol.* **2013**, *84*, S24–S29. [CrossRef]
3. Buset, S.L.; Walter, C.; Friedmann, A.; Weiger, R.; Borgnakke, W.S.; Zitzmann, N.U. Are periodontal diseases really silent? A systematic review of their effect on quality of life. *J. Clin. Periodontol.* **2016**, *43*, 333–344.
4. Herrera, D.; Alonso, B.; Leon, R.; Roldan, S.; Sanz, M. Antimicrobial therapy in periodontitis: The use of systemic antimicrobials against the subgingival biofilm. *J. Clin. Periodontol.* **2008**, *35*, 45–66. [CrossRef]
5. Serrano, J.; Escribano, M.; Roldan, S.; Martin, C.; Herrera, D. Efficacy of adjunctive anti-plaque chemical agents in managing gingivitis: A systematic review and meta-analysis. *J. Clin. Periodontol.* **2015**, *42*, S106–S138. [CrossRef]
6. van Winkelhoff, A.J.; Herrera, D.; Oteo, A.; Sanz, M. Antimicrobial profiles of periodontal pathogens isolated from periodontitis patients in The Netherlands and Spain. *J. Clin. Periodontol.* **2005**, *32*, 893–898. [CrossRef]
7. van Winkelhoff, A.J. Antibiotics in the treatment of peri-implantitis. *Eur. J. Oral Implantol.* **2012**, *5*, S43–S50. [PubMed]
8. Rams, T.E.; Degener, J.E.; van Winkelhoff, A.J. Antibiotic resistance in human peri-implantitis microbiota. *Clin. Oral Implants. Res.* **2014**, *25*, 82–90. [CrossRef] [PubMed]
9. Strydonck, D.A.C.; Slot, D.E.; Van der Velden, U.; Van der Weijden, F. Effect of a chlorhexidine mouthrinse on plaque, gingival inflammation and staining in gingivitis patients: A systematic review. *J. Clin. Periodontol.* **2012**, *39*, 1042–1055. [CrossRef] [PubMed]
10. Cozens, D.; Read, R.C. Anti-adhesion methods as novel therapeutics for bacterial infections. *Expert. Rev. Anti. Infect. Ther.* **2012**, *10*, 1457–1468. [CrossRef]
11. Gutierrez, S.; Moran, A.; Martinez-Blanco, H.; Ferrero, M.A.; Rodriguez-Aparicio, L.B. The Usefulness of Non-Toxic Plant Metabolites in the Control of Bacterial Proliferation. *Probiotics Antimicrob. Proteins* **2017**, *9*, 323–333. [CrossRef] [PubMed]
12. Sánchez, M.C.; Ribeiro-Vidal, H.; Esteban-Fernández, A.; Bartolomé, B.; Figuero, E.; Moreno-Arribas, M.V.; Sanz, M.; Herrera, D. Antimicrobial activity of red wine and oenological extracts against periodontal pathogens in a validated oral biofilm model. *BMC Complement. Altern. Med.* **2019**, *19*, 145. [CrossRef] [PubMed]
13. Pappas, E.; Schaich, K.M. Phytochemicals of cranberries and cranberry products: Characterization, potential health effects, and processing stability. *Crit. Rev. Food Sci. Nutr.* **2009**, *49*, 741–781. [CrossRef] [PubMed]
14. Gyawali, R.; Ibrahim, S.A. Impact of plant derivatives on the growth of foodborne pathogens and the functionality of probiotics. *Appl. Microbiol. Biotechnol.* **2012**, *95*, 29–45. [CrossRef]
15. Jensen, H.D.; Struve, C.; Christensen, S.B.; Krogfelt, K.A. Cranberry juice and combinations of its organic acids are effective against experimental urinary tract infection. *Front. Microbiol.* **2017**, *8*, 542. [CrossRef]

16. Sun, J.; Marais, J.P.; Khoo, C.; LaPlante, K.; Vejborg, R.M.; Givskov, M.; Tolker-Nielsen, T.; Seeram, N.P.; Rowley, D.C. Cranberry (*Vaccinium macrocarpon*) oligosaccharides decrease biofilm formation by uropathogenic *Escherichia coli*. *J. Funct. Foods* **2015**, *17*, 235–242. [CrossRef]

17. Gonzalez de Llano, D.; Liu, H.; Khoo, C.; Moreno-Arribas, M.V.; Bartolome, B. Some new findings regarding the antiadhesive activity of cranberry phenolic compounds and their microbial-derived metabolites against uropathogenic bacteria. *J. Agric. Food. Chem.* **2019**, *67*, 2166–2174. [CrossRef]

18. Maisuria, V.B.; Los Santos, Y.L.; Tufenkji, N.; Deziel, E. Cranberry-derived proanthocyanidins impair virulence and inhibit quorum sensing of *Pseudomonas aeruginosa*. *Sci. Rep.* **2016**, *6*, 30169. [CrossRef]

19. Ulrey, R.K.; Barksdale, S.M.; Zhou, W.; van Hoek, M.L. Cranberry proanthocyanidins have anti-biofilm properties against *Pseudomonas aeruginosa*. *BMC Complement. Altern. Med.* **2014**, *14*, 499. [CrossRef] [PubMed]

20. Yan, X.; Murphy, B.T.; Hammond, G.B.; Vinson, J.A.; Neto, C.C. Antioxidant activities and antitumor screening of extracts from cranberry fruit (*Vaccinium macrocarpon*). *J. Agric. Food. Chem.* **2002**, *50*, 5844–5849. [CrossRef]

21. Eydelnant, I.A.; Tufenkji, N. Cranberry derived proanthocyanidins reduce bacterial adhesion to selected biomaterials. *Langmuir* **2008**, *24*, 10273–10281. [CrossRef] [PubMed]

22. Feghali, K.; Feldman, M.; La, V.D.; Santos, J.; Grenier, D. Cranberry proanthocyanidins: Natural weapons against periodontal diseases. *J. Agric. Food Chem.* **2012**, *60*, 5728–5735. [CrossRef] [PubMed]

23. O'May, C.; Tufenkji, N. The swarming motility of *Pseudomonas aeruginosa* is blocked by cranberry proanthocyanidins and other tannin-containing materials. *Appl. Environ. Microbiol.* **2011**, *77*, 3061–3067. [CrossRef] [PubMed]

24. O'May, C.; Ciobanu, A.; Lam, H.; Tufenkji, N. Tannin derived materials can block swarming motility and enhance biofilm formation in *Pseudomonas aeruginosa*. *Biofouling* **2012**, *28*, 1063–1076. [CrossRef] [PubMed]

25. Grosso, G.; Micek, A.; Godos, J.; Pajak, A.; Sciacca, S.; Galvano, F.; Giovannucci, E.L. Dietary flavonoid and lignan intake and mortality in prospective cohort studies: Systematic review and dose-response meta-analysis. *Am. J. Epidemiol.* **2017**, *185*, 1304–1316. [CrossRef]

26. Xu, H.; Luo, J.; Huang, J.; Wen, Q. Flavonoids intake and risk of type 2 diabetes mellitus: A meta-analysis of prospective cohort studies. *Medicine* **2018**, *97*, e0686. [CrossRef]

27. Neto, C.C.; Amoroso, J.W.; Liberty, A.M. Anticancer activities of cranberry phytochemicals: An update. *Mol. Nutr. Food Res.* **2008**, *5*, S18–S27. [CrossRef]

28. Singh, A.P.; Singh, R.K.; Kim, K.K.; Satyan, K.S.; Nussbaum, R.; Torres, M.; Brand, L.; Vorsa, N. Cranberry proanthocyanidins are cytotoxic to human cancer cells and sensitize platinum-resistant ovarian cancer cells to paraplatin. *Phytother. Res.* **2009**, *23*, 1066–1074. [CrossRef]

29. Bodet, C.; Grenier, D.; Chandad, F.; Ofek, I.; Steinberg, D.; Weiss, E.I. Potential oral health benefits of cranberry. *Crit. Rev. Food. Sci. Nutr.* **2008**, *48*, 672–680. [CrossRef]

30. Philip, N.; Walsh, L.J. Cranberry polyphenols: Natural weapons against dental caries. *Dent. J.* **2019**, *7*, 20. [CrossRef]

31. Bonifait, L.; Grenier, D. Cranberry polyphenols: Potential benefits for dental caries and periodontal disease. *J. Can. Dent. Assoc.* **2010**, *76*, a130. [PubMed]

32. Steinberg, D.; Feldman, M.; Ofek, I.; Weiss, E.I. Cranberry high molecular weight constituents promote *Streptococcus sobrinus* desorption from artificial biofilm. *Int. J. Antimicrob. Agents* **2005**, *25*, 247–251. [CrossRef] [PubMed]

33. Koo, H.; Nino de Guzman, P.; Schobel, B.D.; Vacca Smith, A.V.; Bowen, W.H. Influence of cranberry juice on glucan-mediated processes involved in *Streptococcus mutans* biofilm development. *Caries Res.* **2006**, *40*, 20–27. [CrossRef] [PubMed]

34. Koo, H.; Duarte, S.; Murata, R.M.; Scott-Anne, K.; Gregoire, S.; Watson, G.E.; Singh, A.P.; Vorsa, N. Influence of cranberry proanthocyanidins on formation of biofilms by *Streptococcus mutans* on saliva-coated apatitic surface and on dental caries development *in vivo*. *Caries Res.* **2010**, *44*, 116–126. [CrossRef]

35. Gregoire, S.; Singh, A.P.; Vorsa, N.; Koo, H. Influence of cranberry phenolics on glucan synthesis by glucosyltransferases and *Streptococcus mutans* acidogenicity. *J. Appl. Microbiol.* **2007**, *103*, 1960–1968. [CrossRef]

36. Yamanaka, A.; Kimizuka, R.; Kato, T.; Okuda, K. Inhibitory effects of cranberry juice on attachment of oral streptococci and biofilm formation. *Oral Microbiol. Immunol.* **2004**, *19*, 150–154. [CrossRef]

37. Labrecque, J.; Bodet, C.; Chandad, F.; Grenier, D. Effects of a high-molecular-weight cranberry fraction on growth, biofilm formation and adherence of *Porphyromonas gingivalis*. *J. Antimicrob. Chemother.* **2006**, *58*, 439–443. [CrossRef]

38. Yamanaka, A.; Kouchi, T.; Kasai, K.; Kato, T.; Ishihara, K.; Okuda, K. Inhibitory effect of cranberry polyphenol on biofilm formation and cysteine proteases of *Porphyromonas gingivalis*. *J. Periodontal Res.* **2007**, *42*, 589–592. [CrossRef]

39. Sánchez-Patán, F.; Bartolomé, B.; Martín-Álvarez, P.J.; Anderson, M.; Howell, A.; Monagas, M. Comprehensive assessment of the quality of commercial cranberry products. Phenolic characterization and in vitro bioactivity. *J. Agric. Food Chem.* **2012**, *60*, 3396–3408. [CrossRef]

40. Sanchez, M.C.; Llama-Palacios, A.; Blanc, V.; Leon, R.; Herrera, D.; Sanz, M. Structure, viability and bacterial kinetics of an in vitro biofilm model using six bacteria from the subgingival microbiota. *J. Periodontal Res.* **2011**, *46*, 252–260. [CrossRef]

41. Keijser, J.A.; Verkade, H.; Timmerman, M.F.; Van der Weijden, F.A. Comparison of 2 commercially available chlorhexidine mouthrinses. *J. Periodontol.* **2003**, *74*, 214–218. [CrossRef] [PubMed]

42. Herrera, D.; Roldan, S.; Santacruz, I.; Santos, S.; Masdevall, M.; Sanz, M. Differences in antimicrobial activity of four commercial 0.12% chlorhexidine mouthrinse formulations: An in vitro contact test and salivary bacterial counts study. *J. Clin. Periodontol.* **2003**, *30*, 307–314. [CrossRef] [PubMed]

43. Sekino, S.; Ramberg, P. The effect of a mouth rinse containing phenolic compounds on plaque formation and developing gingivitis. *J Clin. Periodontol.* **2005**, *32*, 1083–1088. [CrossRef] [PubMed]

44. Sanchez, M.C.; Marin, M.J.; Figuero, E.; Llama-Palacios, A.; Leon, R.; Blanc, V.; Herrera, D.; Sanz, M. Quantitative real-time PCR combined with propidium monoazide for the selective quantification of viable periodontal pathogens in an in vitro subgingival biofilm model. *J. Periodontal Res.* **2014**, *49*, 20–28. [CrossRef]

45. Daglia, M. Polyphenols as antimicrobial agents. *Curr. Opin. Biotechnol.* **2012**, *23*, 174–181. [CrossRef]

46. Slobodnikova, L.; Fialova, S.; Rendekova, K.; Kovac, J.; Mucaji, P. Antibiofilm activity of plant polyphenols. *Molecules* **2016**, *21*, 1717. [CrossRef]

47. Nogueira, M.C.; Oyarzabal, O.A.; Gombas, D.E. Inactivation of *Escherichia coli* O157:H7, *Listeria monocytogenes*, and *Salmonella* in cranberry, lemon, and lime juice concentrates. *J. Food. Prot.* **2003**, *66*, 1637–1641. [CrossRef]

48. Wojnicz, D.; Tichaczek-Goska, D.; Korzekwa, K.; Kicia, M.; Hendrich, A.B. Study of the impact of cranberry extract on the virulence factors and biofilm formation by *Enterococcus faecalis* strains isolated from urinary tract infections. *Int. J. Food Sci. Nutr.* **2016**, *67*, 1005–1016. [CrossRef]

49. Cote, J.; Caillet, S.; Doyon, G.; Dussault, D.; Sylvain, J.F.; Lacroix, M. Antimicrobial effect of cranberry juice and extracts. *Food Control* **2011**, *22*, 1413–1418. [CrossRef]

50. La, V.D.; Howell, A.B.; Grenier, D. Anti-*Porphyromonas gingivalis* and Anti-Inflammatory Activities of A-Type Cranberry Proanthocyanidins. *Antimicrob. Agents Chemother.* **2010**, *54*, 1778–1784. [CrossRef]

51. Chrubasik-Hausmann, S.; Vlachojannis, C.; Zimmermann, B.F. Proanthocyanin Content in Cranberry CE Medicinal Products. *Phytother. Res.* **2014**, *28*, 1612–1614. [CrossRef] [PubMed]

52. Costerton, J.W.; Lewandowski, Z.; Caldwell, D.E.; Korber, D.R.; Lappin-Scott, H.M. Microbial biofilms. *Annu. Rev. Microbiol.* **1995**, *49*, 711–745. [CrossRef] [PubMed]

53. Mah, T.F.; O'Toole, G.A. Mechanisms of biofilm resistance to antimicrobial agents. *Trends Microbiol.* **2001**, *9*, 34–39. [CrossRef]

54. Donlan, R.M.; Costerton, J.W. Biofilms: Survival mechanisms of clinically relevant microorganisms. *Clin. Microbiol. Rev.* **2002**, *15*, 167–193. [CrossRef] [PubMed]

55. Davies, D. Understanding biofilm resistance to antibacterial agents. *Nat. Rev. Drug. Discov.* **2003**, *2*, 114–122. [CrossRef] [PubMed]

56. Philip, N.; Bandara, H.; Leishman, S.J.; Walsh, L.J. Inhibitory effects of fruit berry extracts on *Streptococcus mutans* biofilms. *Eur. J. Oral Sci.* **2019**, *127*, 122–129. [CrossRef] [PubMed]

57. Yamanaka-Okada, A.; Sato, E.; Kouchi, T.; Kimizuka, R.; Kato, T.; Okuda, K. Inhibitory effect of cranberry polyphenol on cariogenic bacteria. *Bull Tokyo Dent. Coll.* **2008**, *49*, 107–112. [CrossRef]

58. Polak, D.; Naddaf, R.; Shapira, L.; Weiss, E.I.; Houri-Haddad, Y. Protective potential of non-dialyzable material fraction of cranberry juice on the virulence of *P. gingivalis* and *F. nucleatum* mixed infection. *J. Periodontol.* **2013**, *84*, 1019–1025. [CrossRef]

59. Bodet, C.; Chandad, F.; Grenier, D. Anti-inflammatory activity of a high molecular weight cranberry fraction on macrophages stimulated by lipopolysaccharides from periodontopathogens. *J. Dent. Res.* **2006**, *85*, 235–239. [CrossRef]

60. Mi, H.; Wang, D.; Xue, Y.; Zhang, Z.; Niu, J.; Hong, Y.; Drlica, K.; Zhao, X. Dimethyl sulfoxide protects *escherichia coli* from rapid antimicrobial-mediated killing. *Antimicrob. Agents Chemother.* **2016**, *60*, 5054–5058. [CrossRef]

61. Su, P.W.; Yang, C.H.; Yang, J.F.; Su, P.Y.; Chuang, L.Y. Antibacterial activities and antibacterial mechanism of *Polygonum cuspidatum* extracts against nosocomial drug-resistant pathogens. *Molecules* **2015**, *20*, 11119–11130. [CrossRef] [PubMed]

62. Galvao, J.; Davis, B.; Tilley, M.; Normando, E.; Duchen, M.R.; Cordeiro, M.F. Unexpected low-dose toxicity of the universal solvent DMSO. *FASEB J.* **2014**, *28*, 1317–1330. [CrossRef] [PubMed]

63. Kirkwood, Z.I.; Millar, B.C.; Downey, D.G.; Moore, J.E. Antimicrobial effect of dimethyl sulfoxide and N, N-Dimethylformamide on Mycobacterium abscessus: Implications for antimicrobial susceptibility testing. *Int. J. Mycobacteriol.* **2018**, *7*, 134–136. [PubMed]

64. Weiss, E.I.; Kozlovsky, A.; Steinberg, D.; Lev-Dor, R.; Bar Ness Greenstein, R.; Feldman, M.; Sharon, N.; Ofek, I. A high molecular mass cranberry constituent reduces *mutans* streptococci level in saliva and inhibits in vitro adhesion to hydroxyapatite. *FEMS Microbiol. Lett.* **2004**, *232*, 89–92. [CrossRef]

Article

Improvement of the Flavanol Profile and the Antioxidant Capacity of Chocolate Using a Phenolic Rich Cocoa Powder

Rocío González-Barrio *, Vanesa Nuñez-Gomez, Elena Cienfuegos-Jovellanos, Francisco Javier García-Alonso and Mª Jesús Periago-Castón

Department of Food Technology, Food Science and Nutrition, Faculty of Veterinary Sciences, Regional Campus of International Excellence "Campus Mare Nostrum", Biomedical Research Institute of Murcia (IMIB-Arrixaca-UMU), University Clinical Hospital "Virgen de la Arrixaca", University of Murcia, Espinardo, 30100 Murcia, Spain; vanesa.nunez@um.es (V.N.-G.); elena.cienfuegosjovellanos@gmail.com (E.C.-J.); fjgarcia@um.es (F.J.G.-A.); mjperi@um.es (M.J.P.-C.)
* Correspondence: rgbarrio@um.es; Tel.: +34-868-88-96-41

Received: 12 January 2020; Accepted: 12 February 2020; Published: 14 February 2020

Abstract: Chocolate is made from cocoa, which is rich in (poly)phenols that have a high antioxidant capacity and are associated with the prevention of chronic diseases. In this study, a new production process was evaluated in order to obtain a dark chocolate enriched in (poly)phenols using a cocoa powder with an improved flavanol profile. The antioxidant capacity (Oxygen Radical Absorbance Capacity (ORAC) assay) and the flavanol profile (HPLC-DAD and HPLC-FL) was determined. The analysis of the enriched chocolate showed that the total flavan-3-ols (monomers) content was 4 mg/g representing a 3-fold higher than that quantified in the conventional one. Total levels of dimers (procyanidin B1 and B2) were 2.4-fold higher in the enriched chocolate than in the conventional, with a total content of 6 mg/g. The total flavanol content (flavan-3-ols and procyanidins) in the enriched chocolate was increased by 39% compared to the conventional one which led to a 56% increase in the antioxidant capacity. The new flavanol-enriched dark chocolate is expected to provide greater beneficial effect to consumers. Moreover, the amount of flavonols provided by a single dose (ca. 200 mg per 10 g) would allow the use of a health claim on cardiovascular function, a fact of interest for the cocoa industry.

Keywords: phenolic compounds; HPLC-DAD; fluorescence detection; flavan-3-ols; procyanidins; ORAC; (+)-catechin; (−)-epicatechin; dark chocolate

---

1. Introduction

In recent years, the consumption of cocoa and cocoa products has aroused greater interest due to their beneficial health effects. Cocoa beans and their derivate products are rich in (poly)phenols, which are associated with the prevention of diseases related to oxidative stress, such as cardiovascular diseases, carcinogenic processes, and neurodegenerative diseases [1–3]. However, the beneficial effects of cocoa (poly)phenols depend on the amount consumed, their bioavailability, and the biological activity of the conjugates formed [4]. Cocoa contains several classes of phenolic compounds among which, flavan-3-ols, procyanidins and anthocyanins [5]. Flavanols (flavan-3-ols and procyanidins) are the most studied compounds in cocoa and its derivatives for their beneficial health effects. The monomers (+)-catechin fond (−)-epicatechin and the dimers procyanidin B1 and procyanidin B2 have been identified as the main flavanols in cocoa beans. Among the flavan-3-ols, the major one is (−)-epicatechin, and among the procyanidins, its dimer B2, ranging from 0.4 to 4.5 mg/g DW and 0.3 to 10 mg/g DW, respectively, depending on the geographic origin [6]. At much lower concentrations, (+)-catechin and procyanidin B1 have also been identified in cocoa, ranging from 0.2 to 0.7 mg/g DW and 0.003 to 0.05 mg/g DW, respectively. In addition, trimers and even decamers of flavan-3-ols have also been identified in cocoa and dark chocolate [7,8].

The contents of (poly)phenols in cocoa beans depends on factors, such as variety (genotype) and origin, as well as post-harvest treatments [5,9,10]. The technological process related to the processing of cocoa beans for chocolate manufacturing affect the flavanol profile, both qualitative and quantitative [5,11]. Chocolate-making consists of a multistep process. The cocoa beans are cleaned and separated prior to industrial processing. Afterwards, they are roasted; a critical stage in the final product, since on the one hand guarantees the safety of the product and on the other hand the development of the cocoa flavors. But, it should also be noted that the high of roasting temperatures and time, lead to the loss of bioactive compounds of interest, such as (poly)phenols. Therefore, the optimization of the process to limit the loss of (poly)phenols can be of great importance for cocoa industry that wants to develop a healthy cocoa product, ensuring the safety and quality of the final product. Hence, several studies have been carried out to develop procedures to reduce the loss of phenolic compounds because there are many process that can significantly reduce the (poly)phenol content in cocoa [7,12,13]. It is known that the preservation of polyphenols during cocoa manufacture is important to the beneficial effects on human health associated with the consumption of cocoa and derived products. In a previous study, the process for obtaining a (poly)phenol enriched cocoa powder was described [14]. In this new process, some traditional stages, such as fermentation and roasting, were avoided and a step to inactivate the enzyme polyphenol oxidase was included, which helps preserve the polyphenol content in the raw cocoa bean.

The optimization of the industrial process can increase the percentage of (poly)phenols in the final products, such as cocoa powder, enabling them to be used as functional ingredients [15,16]. Several studies have shown that extracts from different cocoa beans and cocoa liquor have a powerful antioxidant activity, provided by the presence of flavonoids [8,17,18].

Many studies have described the benefits associated with the consumption of cocoa and its derivatives and this has increased the interest in obtaining functional cocoa products. As mentioned above, the benefits for human health depend not only on the amount of (poly)phenols consumed, but also on the bioavailability of the bioactive compounds [19]. In this regard, many studies have reported that the monomers, (+)-catechin and (−)-epicatechin, and the dimer procyanidin B2 are more bioavailable, being more beneficial than other flavanols with a greater molecular size [20]. The aim of this study was to evaluate the improvement in the flavanols profile and in the antioxidant capacity of the dark chocolate when using a (poly)phenol enriched cocoa powder, with the intention of yielding a functional product with potential beneficial effects for human health.

2. Materials and Methods

2.1. Chemical Reagents and Standards

All chemicals were purchased from Sigma Aldrich (St. Louis, MO, USA) while HPLC-grade solvents were purchased from either Scharlab (Barcelona, Spain) or Merck (Darmstadt, Germany). Standards of (+)-catechin and (−)-epicatechin (Sigma Aldrich, St. Louis, MO, USA) and procyanidin B1 and procyanidin B2 (Extrasynthese, Genay, France) were used for quantitative determinations.

2.2. Development of (Poly)Phenol Enriched Dark Chocolate and Conventional Dark Chocolate

Batches of 4 kg of (poly)phenol enriched chocolate and 4 kg of conventional dark chocolate were produced on a pilot scale (Figure 1). For the production of both chocolates, cocoa liquor (58%), sugar (20.5%), cocoa butter (6%), lecithin (0.45%), and vanillin (0.05%) were used, adding 15% (poly)phenol enriched cocoa powder for the enriched chocolate and 15% conventional cocoa powder for the conventional chocolate. The enriched cocoa powder was produced on an industrial scale from unfermented, blanched, and non-roasted cocoa beans (Amazonic-Trinitary variety-CCN51 clone from

Ecuador) using the procedure described by Cienfuegos-Jovellanos et al. [21]. Briefly, a blanching treatment with water (95 °C, 5 min) was applied to fresh cocoa beans, after removal of the pulp. This enriched cocoa cake was defatted by expeller pressing up to a fat content of 11%, and micronized to give a particle size of less than 75 microns. The conventional cocoa powder (Granada brand) was a commercial cocoa powder supplied by Natra Cacao S.L. (Valencia, Spain). It was produced from fermented and dried cocoa beans (Forastero variety from East Africa), being husked, roasted, defatted by hydraulic pressing up to a fat content of 11%, and micronized to a particle size of less than 75 microns. Both cocoa powders, the enriched and the conventional, had a moisture content of 5% and a pH of 5.6. The percentage of cocoa powder (15%) added was because in the sensory test no great differences were found between the enriched chocolate and the conventional one. This was confirmed by an internal tasting panel where a group of volunteers tasted different versions of chocolates with various percentages of enriched cocoa powder, and, apparently, the one with 15% added did not seem different from the standard chocolate and was acceptable in terms of overall sensory quality.

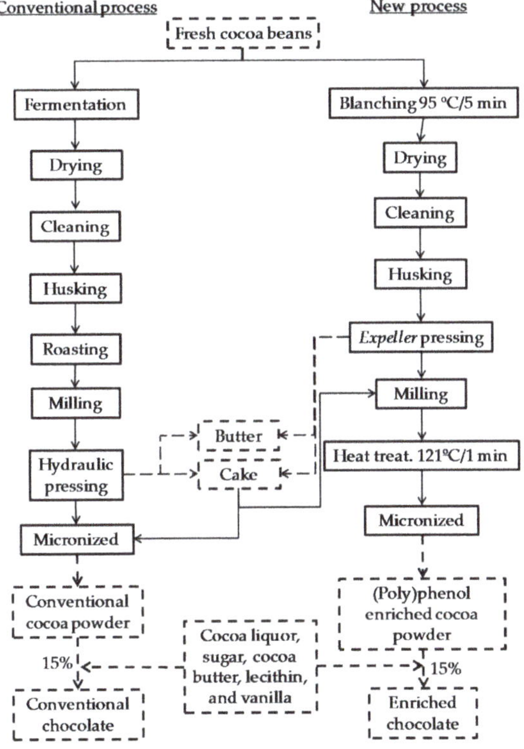

Figure 1. Diagram of the conventional and the new production process of (poly)phenol enriched cocoa powder [21] and chocolate.

The chocolate manufacturing process involved four steps: mixing/kneading, refining, conching, and tempering. In the first step, the ingredients (cocoa liquor, sugar, cocoa powder, and cocoa butter) were mixed together and kneaded, using refining or grinding procedures to provide a smooth chocolate paste. Then, the chocolate was subjected to the conching process, being mechanically kneaded to give it a more complete, homogeneous aroma and improved rheological characteristics. This step was carried out at temperatures between 75 °C and 80 °C, and other components—such as vanillin and lecithin—were added. Finally, the liquid chocolate was tempered, by cooling and heating under controlled conditions, and placed in molds.

2.3. Determination of the Antioxidant Capacity

The antioxidant capacity in cocoa powder and chocolate samples was determined by the Oxygen Radical Absorbance Capacity (ORAC) method, according to Cao et al. [22]. This method is based on the inhibition of the oxidation induced by a peroxy-radical, using a standard with antioxidant capacity as the substrate and a fluorescent probe to measure the signal. Fluorescein was used as the indicator, Trolox as the standard, and 2,2'-azobis(2-amidino-propane) dihydrochloride (AAPH) as the peroxyl radical generator. The assay was carried out using a fluorescent microplate reader (Synergy 2 Multi-Mode Microplate Reader, Biotek, Winooski, VT, USA) and 96-well black microplates equipped with a fluorescence filter having an excitation wavelength of 485 nm and an emission wavelength of 520 nm. For each calibration solution, the blank (0.075 M phosphate buffer, pH 7.0) and the samples were added to the corresponding wells. The plate reader also has an incubator and two injection pumps, which added the fluorescein and the AAPH during the assay; the temperature of the incubator was set to 37 °C. The fluorescence of each well was measured every 60 s for 90 min. The results were calculated using the standard curve provided by the instrument and expressed as µmol of Trolox Equivalents per g (µmol TE/g).

2.4. Analysis of Flavanols by High Performance Liquid Chromatography with UV-Vis Detection (HPLC-DAD)

The analysis of (+)-catechin and (−)-epicatechin monomers, as well as the dimers procyanidin B1 and procyanidin B2, in cocoa powder and chocolate samples was carried out by HPLC-DAD according to the method described by Andrés-Lacueva et al. [23], with some modifications. The chocolate samples were melted at 40 °C before defatting of the sample by the Soxhlet method, using petroleum ether at 55 °C for 24 h, and then drying of the sample at 40 °C. The defatted sample was homogenized in a vortex mixer with 5 mL of distilled water at 100 °C for 1 min. Twenty milliliters of methanol acidified with 0.1% HCl were added and agitated in a vortex for 2 min. Then, the homogenate was centrifuged at 1600*g* for 15 min, at 4 °C. This procedure was repeated twice, and the supernatants were combined and evaporated at 35 °C under vacuum to remove the methanol. The remaining aqueous extract was finally filtered through a 0.45-µm PTFE filter and analyzed by HPLC-DAD. The Agilent 1100 HPLC system was fitted with a quaternary pump, a degasser, a thermostatted column support, an autosampler, and a serial diode detector (Agilent Technologies, Waldbronn, GermanySpain). The separation of the different monomers and dimers was performed using a C18 Zorbax Eclipse XDB reverse phase column (150 × 2.1 mm, i.d. 5 µm) (Agilent Technologies, Spain) thermostatted at 35 °C. The mobile phases used were 0.1% aqueous formic acid (solvent A) and acetonitrile (B), at a flow rate of 0.6 mL/min. Elution began with a gradient from 4 to 10% B in 25 min, followed by a gradient to 13% B at 30 min, to 15% B at 33 min, and to 50% B at 35 min, followed by washing and then a return to the initial conditions at 45 min. Chromatograms were recorded at 280 nm. The identification of flavanols was performed by comparison of their retention times with those of the authentic standards of (+)-catechin hydrate (−)-epicatechin, procyanidin B1 and procyanidin B2. Quantification was based on calibration curves constructed using 5 to 25 ppm (+)-catechin, 20 to 100 ppm (−)-epicatechin, 1 to 10 ppm procyanidin B1, and 5 to 50 ppm procyanidin B2.

2.5. Analysis of Flavan-3-ols and Procyanidins by High Performance Liquid Chromatography and Fluorescence Detection (HPLC-FL)

The analysis of procyanidins in the cocoa powder and chocolate samples by HPLC-FL was carried out according to the method described by Gu et al. [8]. This included defatting of the sample (2 g) using 45 mL of n-hexane, followed by centrifugation and evaporation. After that, the extraction of the procyanidins was performed according to the methodology of Gu et al. [24]. Briefly, defatted samples were homogenized in a vortex mixer with acetone/water/acetic acid (70/29.5/0.5, *v/v/v*), sonicated at 37 °C for 15 min, and finally centrifuged at 3000 *g* for 5 min at 4 °C. The extraction process was performed twice more, each time adding 15 mL of the acetone/water/acetic acid (70/29.5/0.5, *v/v/v*) to the remnant. Finally, the supernatants obtained in each extraction stage were combined and the organic phase was removed under vacuum. The aqueous phase remaining was shaken in an ultrasonic bath and purified using a vacuum-equipped solid phase extraction unit (SPE) (Merck, Darmstadt, Germany). A lipophilic filler of Sephadex LH-20 (Scharlab, Spain) was used as the solid phase. For each sample, 3 g of Sephadex LH-20 were packed into a column (6 × 1.5 cm) and preconditioned with 15 mL of methanol/water (30/70, *v/v*) overnight before use. The sample was passed through sorbent and collected. Once the procyanidins had been retained in the adsorbent, the column was washed with 40 mL of a mixture of methanol/water (30/70, *v/v*) to remove the sugars and other phenols. The procyanidins were then recovered with 80 mL of acetone/water (70/30, *v/v*) until the cartridge was completely clean. Finally, the acetone was evaporated at 45 °C under vacuum. The aqueous extract obtained was lyophilized and dissolved in 10 mL of acetone/water/acetic acid (70/29.5/0.5, *v/v/v*), before being filtered and injected into the HPLC-FL system. The different procyanidin fractions were separated, according to their degree of polymerization, in the same Agilent 1100 HPLC system described above. Separation of the different fractions was achieved on a C-8 normal phase column, Luna Silica 100 Å (150 × 4.6 mm, i.d. 5 μm) (Phenomenex, Madrid, Spain). The mobile phase used was a mixture of dichloromethane (A), methanol (B), and glacial acetic acid/water (50/50, *v/v*) (C) at a flow rate of 0.5 mL/min. Elution began with a linear gradient from 14 to 28% B in 30 min, followed by a linear gradient from 28 to 39% B at 45 min, and from 39 to 86% B at 50 min. From 50 to 55 min an isocratic gradient of 86% B was used, followed by washing and then a return to the initial conditions at 70 min. A presence of 4% C remained constant throughout the elution. A fluorescence detector, set at an excitation wavelength of 276 nm and an emission wavelength of 316 nm, was used for detection. Due to the lack of commercially available standards, the quantification of the different procyanidin fractions was performed using previously published relative response factors for each one [25]. Briefly, the response factor of (−)-epicatechin was measured based on a calibration curve constructed using a commercial standard. Then, the response factors of the different procyanidin fractions were estimated using the corresponding relative response factors (0.65 for dimers, 0.69 for trimers, 0.61 for tetramers, 0.58 for pentamers, 0.45 for hexamers, 0.62 for heptamers, 0.52 for octamers, 0.36 for nonamers, 0.56 for decamers, and 0.45 for >decamers), which were used to calculate the concentration (mg/g) of each fraction.

2.6. Statistical Analysis

All analytical parameters were determined in triplicate for each sample, except the procyanidins analysis by HPLC-FL which was determined in duplicated. The data were expressed as means ± standard deviations (SD). Independent-samples T-tests were applied to determine the differences between means for all analyzed parameters ($p < 0.05$). The statistical analysis was carried out using GraphPad Prism version 6.02 for Windows, GraphPad Software (La Jolla, CA, USA).

3. Results

3.1. Antioxidant Capacity

The ORAC method gave values of 686 μmol TE/g for the conventional cocoa powder and 2861 μmol TE/g (more than 4-times higher) for the (poly)phenol enriched cocoa powder. A value of 412 μmol TE/g was obtained for the conventional dark chocolate and a value of 641 μmol TE/g (1.6-times higher) was obtained for the (poly)phenol enriched chocolate (Figure 2).

3.2. Flavanol Analysis by HPLC-DAD

The qualitative and quantitative profiles of cocoa powder and dark chocolate (enriched and conventional) were analyzed by HPLC-DAD. The chromatogram obtained for the dark (poly)phenols enriched chocolate (Figure 3) provides evidence that in all cases the separation and quantification of (−)-epicatechin and (+)-catechin monomers and procyanidin B2 and B1 polymers were achieved.

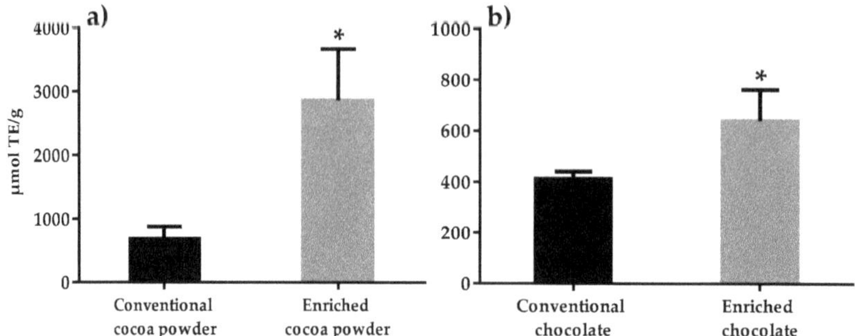

Figure 2. Oxygen Radical Absorbance Capacity (ORAC) antioxidant capacity of conventional and enriched cocoa powder (**a**) and chocolate (**b**). * Indicates significant differences at $p < 0.05$.

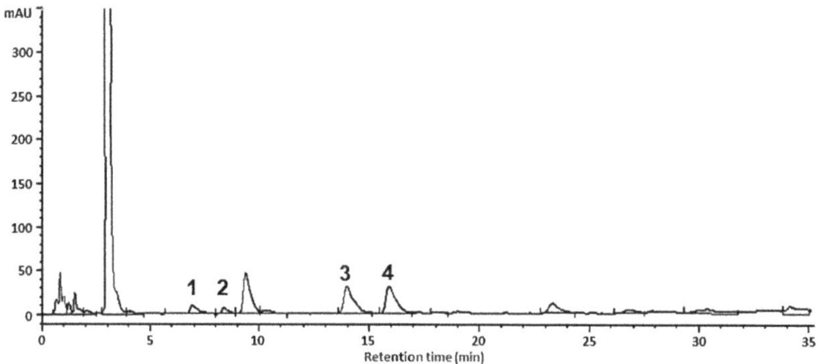

Figure 3. HPLC-DAD chromatogram at 280 nm of dark (poly)phenol enriched chocolate. 1: Procyanidin B1; 2: (+)-catechin; 3: procyanidin B2; 4: (−)-epicatechin.

The results obtained for the cocoa powder and chocolate (conventional and enriched) are shown in Table 1. In both samples (cocoa powder and chocolate) the main flavanols found were (−)-epicatechin and procyanidin B2.

Table 1. Flavanol contents (mg/g) in the conventional and enriched dark chocolate analyzed by reverse phase HPLC-DAD [1].

Flavanols	Conventional Cocoa Powder	Enriched Cocoa Powder	Conventional Chocolate	Enriched Chocolate
(+)-catechin	0.5 ± 0.1	3.4 ± 0.0 *	0.3 ± 0.0	0.6 ± 0.0 ***
(−)-epicatechin	1.9 ± 0.5	24.9 ± 0.0 ***	1.1 ± 0.1	3.4 ± 0.2 ***
Procyanidin B1	0.1 ± 0.0	2.1 ± 0.1 ***	1.2 ± 0.1	1.8 ± 0.1 **
Procyanidin B2	1.0 ± 0.2	14.2 ± 0.3 ***	1.3 ± 0.2	4.2 ± 0.1 ***
Total	3.5 ± 0.8	44.6 ± 0.3 ***	3.9 ± 0.1	10.0 ± 0.1 ***

[1] The values are the average of three replicates ($n = 3$) ± SD. *, **, *** Indicates significant differences at $p < 0.05$, $p < 0.01$ and $p < 0.001$, respectively, between conventional and enriched for each products (cocoa powder and chocolate).

Total monomers content was 12-fold higher in the enriched cocoa powder than that quantified in the conventional one, while the total dimers content was 15-fold higher. In the enriched chocolate, the total monomers content was 3-fold higher than in the conventional, while the total dimers content was 2.4-fold higher. Moreover, the total content of monomers and dimers was 3.9 mg/g for the conventional chocolate and 10.0 mg/g (2.6-times greater) for the enriched chocolate (Table 1).

3.3. Procyanidins Analysis by HPLC-FL

Due to the absence of standards for some of the different fractions of procyanidins, their quantification was performed using relative response factors previously published for chocolate [25]. Figure 4 show the chromatograms obtained for the (poly)phenol enriched chocolate. The method used allowed the separation of the different fractions, from monomers to decamers; the procyanidins of a higher degree of polymerization (>10) were quantified at the end of the chromatogram as a single peak.

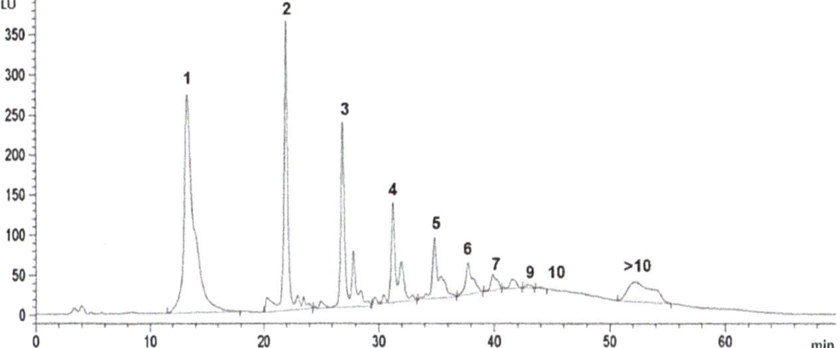

Figure 4. HPLC-FL chromatogram of (poly)phenol enriched cocoa (poly)phenol enriched chocolate (b) 1: monomers; 2: dimers; 3: trimers; 4: tetramers; 5: pentamers; 6: hexamers; 7: heptamers; 8: octamers; 9: nonamers; 10: decamers; >10: procyanidins with a degree of polymerization of more than 10 units.

Table 2 shows the flavanol contents, covering the range from monomers to polymers, in the conventional and (poly)phenol enriched cocoa powder and in the conventional and enriched chocolate. In both the (poly)phenol enriched cocoa powder and enriched chocolate, the monomers were the main fraction, representing 23% of the total flavanols. However, in the conventional cocoa powder and conventional chocolate, the main fraction was comprised of the procyanidins with a degree of polymerization of more than 10 units (>10), this fraction representing 37% and 22%, respectively, of the total flavanols. Overall, the content of procyanidins in the enriched cocoa powder was 4.7-fold greater than in the conventional cocoa powder, and in the enriched chocolate it was 1.4-fold greater than in the conventional chocolate.

Table 2. Flavanol contents (mg/g) of the conventional and enriched cocoa powder and chocolate analyzed by normal phase HPLC-FL [a].

Flavanols	Conventional Cocoa Powder	Enriched Cocoa Powder	Conventional Chocolate	Enriched Chocolate
Monomers	2.70 ± 0.20	17.70 ± 0.10 ***	3.20 ± 0.09	4.70 ± 0.10 **
Dimers	3.00 ± 0.20	16.70 ± 1.40 **	2.90 ± 0.20	4.20 ± 0.30 *
Trimers	2.00 ± 0.20	13.00 ± 0.90 **	2.00 ± 0.01	3.20 ± 0.20 *
Tetramers	1.30 ± 0.08	10.90 ± 0.07 ***	1.50 ± 0.02	2.30 ± 0.10 **
Pentamers	0.77 ± 0.00	7.00 ± 0.06 ***	0.97 ± 0.01	1.50 ± 0.00 ***
Hexamers	0.51 ± 0.00	5.00 ± 0.40 **	0.66 ± 0.01	1.00 ± 0.02 **
Heptamers	0.13 ± 0.00	1.70 ± 0.30 *	0.20 ± 0.01	0.30 ± 0.03 *
Octamers	0.12 ± 0.02	0.90 ± 0.10 **	0.15 ± 0.01	0.20 ± 0.04
Nonamers	0.04 ± 0.01	0.47 ± 0.02 **	0.08 ± 0.02	0.14 ± 0.02
Decamers	nd	0.10 ± 0.10	nd	0.02 ± 0.01
>Decamers	6.20 ± 0.90	4.57 ± 0.19	3.30 ± 0.30	3.20 ± 1.10
Total	16.8 ± 1.6	78.0 ± 1.5 *	15.0 ± 0.5	20.8 ± 0.9 *

[a] The values are the average of two replicates ($n = 2$) ± SD. nd (not detected). *, **, *** Indicates significant differences at $p < 0.05$, $p < 0.01$, and $p < 0.001$, respectively, between conventional and enriched for each product (cocoa powder and chocolate).

4. Discussion

A dark chocolate rich in (poly)phenols has been produced using an enriched cocoa powder, whose new production process reduces the losses of bioactive compounds, allowing these compounds to be preserved during chocolate production. The difference between the enriched and conventional cocoa powders lies in the processing, since the enriched cocoa powder was obtained from a new process in which some stages of the traditional process were replaced, such as fermentation, grain roasting, and defatting by hydraulic pressing. Conversely, other common processes used in the food industry were incorporated to minimize the loss of phenolic compounds, such as scalding to inactivate the enzyme (poly)phenol oxidase (PPO), deffating by expellers at low and controlled temperatures, and heat treatment by steam currents in an autoclave (Figure 1). The dose of cocoa powder added (15%) for the manufacturing of the chocolate samples was chosen on the grounds of a sensory test, in order to obtain no perceivable differences between the enriched chocolate and the conventional one. In spite of the astringency and the purple color of the enriched cocoa powder, this dose did not change the organoleptic properties of the enriched chocolate. Moreover, the enriched cocoa powder had a very plain profile, with no cocoa taste or aroma, since it had not undergone the roasting process; therefore, the precursors of the cocoa aroma and flavor were not developed since it was not fermented either.

The antioxidant capacity of the enriched cocoa powder used for the production of the dark chocolate rich in (poly)phenols was 2861 μmol TE/g representing 317% higher than that of the conventional cocoa powder. This shows the enhanced value of this (poly)phenol enriched cocoa powder. The antioxidant capacity of the enriched chocolate was 641 μmol TE/g representing 56% higher than that of the conventional chocolate, but both values are significantly higher than the corresponding values found in various fruit powders—acai (400 μmol TE/g), blueberry (260 μmol TE/g), cranberry (310 μmol TE/g), and pomegranate (190 μmol TE/g)—and in natural, non-alkalized cocoa powder (634 μmol TE/g) [17].

Taking into account that the chocolates were elaborated with 15% cocoa powder (enriched or conventional), theoretically, the antioxidant capacity should be 103 μmol TE/g for the conventional chocolate and 429 μmol TE/g for the enriched one. However, the values obtained were 4-fold higher for the conventional chocolate (412 μmol TE/g) and 1.5-fold higher for the enriched one (641 μmol TE/g). The differences observed between the hypothetical and the analytical results could be explained, in part, by the presence of other bioactive compounds in the cocoa liquor added (58%) during the chocolate manufacturing [5], which could increase the antioxidant capacity in the chocolate. In addition, the results obtained demonstrate that the enriched powder is less stable than the conventional one during chocolate processing. However, the antioxidant capacity of the enriched chocolate formulation, including 15% (poly)phenols enriched cocoa powder, was increased by 56% compared to the chocolate formulated with 15% conventional cocoa powder.

To achieve beneficial effects on human health, the daily recommended intake of antioxidants is 3000–3600 μmol TE [26]. However, the recommended five pieces of fruit and vegetables per day do not reach 50% of this value, providing between 1200 and 1640 μmol TE [26]. For this reason, the consumption of functional products enriched in (poly)phenols could be a strategy to increase the daily intake of antioxidants. Specifically, a 10-g portion of the enriched dark chocolate described here will provide a high level of antioxidants, more than 6000 μmol TE.

The analysis of the enriched chocolate measured by HPLC-DAD, showed that the total flavan-3-ols content was 4 mg/g representing a 186% higher than that of the conventional chocolate. Moreover, the content in enriched chocolate was also higher than that determined by other authors in dark chocolate with values ranging to 0.24–0.45 mg/g [27]. The results showing that, for chocolate, (−)-epicatechin was the second most abundant of the flavanols analyzed agree with other work reporting that one of the main compounds in cocoa beans is (−)-epicatechin [28]. In this study, the sum of the monomers, (−)-epicatechin and (+)-catechin, in the enriched chocolate (4.0 mg/g) was higher compared to the conventional one (1.4 mg/g), and was higher than that obtained by other authors for a dark chocolate (1.9 mg/g) [8]. These results show that, in the new functional product, there is an increase in the more bioavailable flavanols and therefore in the health benefits of the dark (poly)phenol enriched chocolate, compared to the conventional chocolate. This is because (+)-catechin, (−)-epicatechin and procyanidin B2, which represent 43% of the total procyanidins, are more bioavailable than the other compounds, as some authors have shown, since their absorption in the gastrointestinal tract is higher [20,29–32].

The total flavanol content, covering the range from monomers to polymers, of the enriched cocoa powder was 78 mg/g representing 364% higher than that obtained for the conventional cocoa powder in this study and 90% higher than that obtained for a conventional cocoa powder by other authors who reported a value of 41 mg/g [8]. The mean contents quantified for the flavan-3-ols and procyanidins in this study are similar to the values published for a cocoa powder that was not affected by post-harvest variables [12,33]. Similarly, it should be noted that the total procyanidins content quantified in the (poly)phenol enriched cocoa powder was 1.3-fold higher than that quantified by Kealey et al. [12] in a cocoa powder obtained from fresh unfermented beans, lyophilized and defatted by Soxhlet, and 2.8-fold greater than that quantified by Misnawi et al. [33] in a cocoa powder obtained from fresh unfermented beans partially defatted using an expeller.

Gu et al. [8], using a similar method of extraction, purification with Sephadex LH20, and normal phase analysis using HPLC-ESI/MS with a fluorescence detector, reported values of total procyanidins from 8.5 to 20 mg/g in conventional dark chocolate samples - very similar to the 15 mg/g for conventional chocolate and the 21 mg/g in enriched chocolate found in this study. In addition, the total procyanidin value of 21 mg/g in cocoa powder samples recorded by Miller et al. [34] is close to that found here for conventional cocoa powder (17 mg/g). However, the (poly)phenol enriched chocolate developed in the present study has also been characterized by values of low-molecular-weight procyanidins higher than previously published data ranging values to 1.14–1.77 mg/g [8,35]. The sum of the classes ranging from monomers to hexamers found in the enriched chocolate was 17 mg/g, compared to 10 mg/g found by Gu et al. [8] in a dark chocolate and 11 mg/g found in this work for conventional chocolate. Theoretically, as both chocolates (enriched and conventional) were elaborated with 15% cocoa powder, the total flavanol content should be 11.7 mg/g for the enriched chocolate and 2.5 mg/g for the conventional, 2-fold and 12-fold higher than that quantified for the enriched chocolate and for the conventional one, respectively. These results could be explained, in part, by the presence of flavanols in the cocoa liquor added (58%) during the chocolate manufacturing.

5. Conclusions

Our results describing the formulation of a new dark chocolate enriched with (−)-epicatechin and procyanidin B2 was successfully achieved, as well as a notable enrichment of the oligomeric procyanidin fractions (from monomers to hexamers), when compared with a conventional dark chocolate. These results are of current interest to both large food companies and health professionals. The manufacturing of the enriched dark chocolate following the procedures described above would allow the use of a health claim related to cocoa flavanols. According to the Commission Regulation (EU) 2015/539 of 31 March 2015 [36], the new enriched chocolate could have the claim '*Cocoa flavanols help maintain the elasticity of blood vessels, which contributes to normal blood flow*'. In order to obtain this beneficial effect, 10 g of the new enriched dark chocolate should be consumed daily, which would provide more than 200 mg of total flavanols (flavan-3-ols and procyanidins ranging from dimers to decamers). In addition, the new dark chocolate formulation, including 15% (poly)phenols enriched cocoa powder, increased greatly the antioxidant properties compared to a conventional dark chocolate, while maintaining the organoleptic properties unchanged. However, a future consumer preference study could be of great interest to determine the acceptability of the new functional chocolate.

Author Contributions: Conceptualization, R.G.-B., E.C.-J. and M.J.P.-C.; Formal analysis, V.N.-G. and E.C.-J.; Methodology, R.G.-B. and E.C.-J.; Writing—original draft, R.G.-B. and V.N.-G.; Writing—review & editing, R.G.-B., E.C.-J., F.J.G.-A. and M.J.P.-C. All authors have read and agreed to the published version of the manuscript.

Funding: This research received no external funding.

Acknowledgments: The authors would like to thank F.A. Tomás-Barberán from CEBAS-CSIC (Murcia, Spain) for his technical advice regarding the analysis of flavan-3-ols and procyanidins. All cocoa powder and chocolate samples were kindly provided by Natraceutical Group, Valencia, Spain.

Conflicts of Interest: The authors declare no conflict of interest.

References

1. Martín, M.A.; Goya, L.; Ramos, S. Potential for preventive effects of cocoa and cocoa polyphenols in cancer. *Food Chem. Toxicol.* **2013**, *56*, 336–351. [CrossRef]
2. Crews, W.D., Jr.; Harrison, D.W.; Wright, J.W. A double-blind, placebo controlled, randomized trial of the effects of dark chocolate and cocoa on variables associated with neuropsychological functioning and cardiovascular health: Clinical findings from a sample of healthy, cognitively intact older adults. *Am. J. Clin. Nutr.* **2008**, *7*, 872–880.
3. Andújar, I.; Recio, M.C.; Giner, R.M.; Ríos, J.L. Cocoa polyphenols and their potential benefits for human health. *Oxid. Med. Cell Longev.* **2012**, *2012*, 1–23. [CrossRef]
4. Roura, E.; Andrés-Lacueva, C.; Estruch, R.; Mata-Bilbao, M.L.; Izquierdo-Pulido, M.; Waterhouse, A.L.; Lamuela-Raventós, R.M. Milk does not affect the bioavailability of cocoa powder flavonoid in healthy human. *Ann. Nutr. Metab.* **2007**, *51*, 493–498. [CrossRef]
5. Urbańska, B.; Derewiaka, D.; Lenart, A.; Kowalska, J. Changes in the composition and content of polyphenols in chocolate resulting from pre-treatment method of cocoa beans and technological process. *Eur. Food Res. Technol.* **2019**, *245*, 2101–2112. [CrossRef]
6. Bordiga, M.; Locatelli, M.; Travaglia, F.; Coisson, J.D.; Mazza, G.; Arlorio, M. Evaluation of the effect of processing on cocoa polyphenols: Antiradical activity, anthocyanins and procyanidins profiling from raw beans to chocolate. *Int. J. Food Sci. Tech.* **2015**, *50*, 840–848. [CrossRef]
7. Kelm, M.A.; Johnson, C.; Robbins, R.J.; Hammerstone, J.F.; Schmitz, H.H. High-performance liquid chromatography separation and purification of cacao (*Theobroma cacao* L.) procyanidins according to degree of polymerization using a diol stationary phase. *J. Agric. Food Chem.* **2006**, *54*, 1571–1576. [CrossRef] [PubMed]
8. Gu, L.; House, S.E.; Wu, X.; Ou, B.; Prior, R.L. Procyanidin and catechin contents and antioxidant capacity of cocoa and chocolate products. *J. Agric. Food Chem.* **2006**, *54*, 4057–4061. [CrossRef] [PubMed]
9. Othman, A.; Jalil, A.M.; Weng, K.K.; Ismail, A.; Ghani, N.A.; Adenan, I. Epicatechin content and antioxidant capacity of cocoa beans from four different countries. *Afr. J. Biotechnol.* **2010**, *9*, 1052–1059.
10. Afoakwa, E.O.; Quao, J.; Budu, A.S.; Takrama, J.F.; Saalia, F.K. Changes in total polyphenols, o-diphenols and anthocyanin concentrations during fermentation of pulp pre-conditioned cocoa (Theobroma cacao) beans. *Int. Food Res. J.* **2012**, *19*, 1071–1077.
11. Andrés-Lacueva, C.; Monagas, M.; Khan, N.; Izquierdo-Pulido, M.; Urpi-Sarda, M.; Permanyer, J.; Lamuela-Raventos, R.M. Flavanol and Flavonol Contents of Cocoa Powder Products: Influence of the Manufacturing Process. *J. Agric. Food Chem.* **2008**, *56*, 3111–3117. [CrossRef] [PubMed]
12. Kealey, K.S.; Snyder, R.M.; Romanczyk, L.J.; Geyer, H.M.; Myers, M.E.; Withcare, E.J.; Hammerstone, J.F.; Schmitz, H.H. Cocoa Components, Edible Products Having Enhanced Polyphenol Content, Methods of Making Same and Medical Uses. U.S. Patent 6,312,753; Mars Incorporated, USA, 1998.
13. Zhong, J.L.; Muhammad, N.; Gu, Y.C.; Yan, W.D. A simple and efficient method for enrichment of cocoa polyphenols from cocoa bean husks with macroporus resins following a scale-up separation. *J. Food Eng.* **2019**, *243*, 82–88. [CrossRef]
14. Schinella, G.; Mosca, S.; Cienfuegos-Jovellanos, E.; Pasamar, M.A.; Muguerza, B.; Ramón, D.; Ríos, J.L. Antioxidant properties of polyphenol-rich cocoa products industrially processed. *Food Res. Int.* **2010**, *43*, 1614–1623. [CrossRef]
15. Cienfuegos-Jovellanos, E.; Quiñones, M.M.; Muguerza, B.; Moulay, L.; Miguel, M.; Aleixandre, A. Antihypertensive effect of a polyphenol-rich cocoa powder industrially processed to preserve the original flavonoids of the cocoa beans. *J. Agric. Food Chem.* **2009**, *57*, 57, 6156–6162, 9169–9180. [CrossRef] [PubMed]

16. Ostertag, L.M.; Kroon, P.A.; Wood, S.; Horgan, G.W.; Cienfuegos-Jovellanos, E.; Saha, S.; Duthie, G.G.; de Roos, B. Flavan-3-ol-enriched dark chocolate and white chocolate improve acute measures of platelet function in a gender-specific way—A randomized-controlled human intervention trial. *Mol. Nutr. Food Res.* **2013**, *57*, 191–202. [CrossRef]

17. Crozier, S.J.; Preston, A.G.; Hurst, J.W.; Payne, M.J.; Mann, J.; Hainly, L.; Miller, D. Cacao seeds are a "Super Fruit": A comparative analysis of various fruit powders and products. *Chem. Cent. J.* **2011**, *5*, 1–6. [CrossRef]

18. Miller, K.B.; Stuart, D.A.; Smith, N.L.; Lee, C.Y.; Mc Hale, N.L.; Flanagan, J.A.; Ou, B.; Hurst, W.J. Antioxidant activity and polyphenol and procyanidin contents of selected commercially available cocoa-containing and chocolate products in the United States. *J. Agric. Food Chem.* **2006**, *54*, 4062–4068. [CrossRef]

19. Lamuela-Raventós, R.M.; Romero-Pérez, A.I.; Andrés-Lacueva, C.; Tornero, A. Review: Health effects of cocoa flavonoids. *Int. J. Food Sci. Tech.* **2005**, *11*, 159–176. [CrossRef]

20. Tomás-Barberán, F.A.; Cienfuegos-Jovellanos, E.; Marín, A.; Muguerza, B.; Gil-Izquierdo, A.; Cerdá, B.; Zafrilla, P.; Morillas, J.; Mulero, J.; Ibarra, A.; et al. A new process to develop a cocoa powder with higher flavonoid monomer content and enhanced bioavailability in healthy humans. *J. Agric. Food Chem.* **2007**, *55*, 3926–3935. [CrossRef]

21. Cienfuegos-Jovellanos, E.; Pasamar, M.A.; Fritz, J.; Ramón, D.; Castilla, Y. *Method for Obtaining Polyphenol-Rich Cocoa Powder with a Low Fat Content and Cocoa thus Obtained*; Patent Cooperation Treaty (PCT) WO 2007/096449A1; Natraceutical Industrial: Valencia, Spain, 2007.

22. Cao, G.; Alessio, H.M.; Cutler, R.G. Oxygen-radical absorbance capacity assay for antioxidants. *Free Radical. Biol. Med.* **1993**, *14*, 301–311. [CrossRef]

23. Andrés-Lacueva, C.; Lamuela-Raventós, R.M.; Jáuregui, O.; Casals, I.; Izquierdo-Pulido, M.; Permayer, J. An LC method for the analysis of cocoa phenolics. *LC GC Eur.* **2000**, *13*, 902–905.

24. Gu, L.; Kelm, M.; Hammerstone, J.F.; Beecher, G.; Cunningham, D.; Vannozzi, S.; Prior, R.L. Fractionation of polymeric procyanidins from lowbush blueberry and quantification of procyanidins in selected foods with an optimized normal-phase HPLC-MS fluorescent detection method. *J. Agr. Food Chem.* **2002**, *50*, 4852–4860. [CrossRef]

25. Prior, R.L.; Gu, L. Ocurrence and biological significance of proanthocyanidins in the American diet. *Phytochemistry* **2005**, *66*, 2264–2280. [CrossRef] [PubMed]

26. Prior, R.L.; Cao, G. Antioxidant phytochemicals in fruits and vegetables: Diet and health implications. *Hortscience* **2000**, *35*, 588–592. [CrossRef]

27. Todorovic, V.; Radojcic, I.; Todorovic, Z.; Jankovic, G.; Dodevska, M.; Sobajic, S. Polyphenols, methylxanthines, and antioxidant capacity of chocolates produced in Serbia. *J. Food Compos. Anal.* **2015**, *41*, 137–143. [CrossRef]

28. Rigaud, J.; Escribano-Bailón, M.; Prieur, C.; Souquet, J.M. Normal-phase high-performance liquid chromatographic separation of procyanidins from cacao beans and grape seeds. *J. Chromatogr. A* **1993**, *654*, 255–260. [CrossRef]

29. Richelle, M.; Tavazzi, I.; Enslen, M.; Offord, E.A. Plasma kinetics in man of epicatechin from black chocolate. *Eur. J. Clin. Nutr.* **1999**, *53*, 22–26. [CrossRef]

30. Rein, D.; Lotio, S.; Holt, R.R.; Keen, C.L.; Schmitz, H.H.; Fraga, C.G. Epicatechin in human plasma: In vivo determination and effect of chocolate consumption on a plasma oxidation status. *J. Nutr.* **2000**, *130*, 2109S–2114S. [CrossRef]

31. Holt, R.R.; Lazarus, S.A.; Cameron Sullards, M.; Zhu, Q.Y.; Schramm, D.D.; Hammerstone, J.F.; Fraga, C.G.; Schmitz, H.H.; Keen, C.L. Procyanidin dimer B2 [epicatechin-(4b-8)-epicatechin] inhuman plasma after the consumption of a flavanol-richcocoa. *Am. J. Clin. Nutr.* **2002**, *76*, 798–804. [CrossRef]

32. Andújar, I.; Recio, M.C.; Giner, R.M.; Cienfuegos-Jovellanos, E.; Laghi, S.; Muguerza, B.; Ríos, J.L. Inhibition of ulcerative colitis in mice after oral administration of a polyphenol-enriched cocoa extract is mediated by the inhibition of STAT1 and STAT3 phosphorylation in colon cells. *J. Agric. Food Chem.* **2011**, *59*, 6474–6483. [CrossRef]

33. Misnawi Jinap, S.; Jamilah, B.; Nazamid, S. Effects of incubation and polyphenol oxidase enrichment on colour, fermentation index, procyanidins and astringency of unfermented and partly fermented cocoa beans. *Int. J. Food Sci. Tech.* **2003**, *38*, 285–295. [CrossRef]

34. Miller, K.B.; Hurst, W.J.; Flannigan, N.; Ou, B.; Lee, C.Y.; Smith, N.; Stuart, D. Survey of commercially available chocolate-and cocoa-containing products in the United States. 2. Comparison of flavan-3-ol content with nonfat cocoa solids, total polyphenols, and percent cacao. *J. Agric. Food Chem.* **2009**, *57*, 9169–9180. [CrossRef] [PubMed]

35. Pedan, V.; Fischer, N.; Bernath, K.; Hühn, T.; Rohn, S. Determination of oligomeric proanthocyanidins and their antioxidant capacity from different chocolate manufacturing stages using the NP-HPLC-online-DPPH methodology. *Food Chem.* **2017**, *214*, 523–532. [CrossRef] [PubMed]

36. European Union Law (EUR-Lex). Available online: https://eur-lex.europa.eu/legal-content/EN/TXT/PDF/?uri=CELEX:32015R0539&from=ES (accessed on 10 February 2020).

 *foods*

Article

Alternative Sweeteners Modify the Urinary Excretion of Flavanones Metabolites Ingested through a New Maqui-Berry Beverage

Vicente Agulló [1], Raúl Domínguez-Perles [1,*], Diego A. Moreno [1], Pilar Zafrilla [2] and Cristina García-Viguera [1]

1   Group of Quality, Safety and Bioactivity of Plant Foods, Department of Food Science and Technology, Phytochemistry and Healthy Foods Lab, (CEBAS-CSIC), University Campus of Espinardo 25, 30100 Murcia, Spain; vagullo@cebas.csic.es (V.A.); dmoreno@cebas.csic.es (D.A.M.); cgviguera@cebas.csic.es (C.G.-V.)
2   Department of Pharmacy, Faculty of Health Sciences, Universidad Católica San Antonio de Murcia (UCAM), Campus de los Jerónimos, 30107 Guadalupe, Murcia, Spain; mpzafrilla@ucam.edu
*   Correspondence: rdperles@cebas.csic.com; Tel.: +34-968-396-200

Received: 1 November 2019; Accepted: 26 December 2019; Published: 3 January 2020

Abstract: Dietary sugar has been largely related to the onset of metabolic diseases such as type 2 diabetes and obesity, among others. The growing awareness on the close relationship between the dietary habits and this health disturbance has encouraged the development of new beverages using alternative sweeteners that could contribute to combat the above referred pathophysiological disorders. To gain further insight into this issue, the present work, upon an acute dietary intervention, evaluated the urinary excretion of flavanones ingested through polyphenols-rich beverages composed of maqui berry and citrus, with the aim of establishing the highest urinary excretion rate and metabolite profiles. The functional beverages evaluated were supplemented with a range of sweeteners including sucrose (natural and high caloric), stevia (natural and non-caloric), and sucralose (artificial and non-caloric) as an approach that would allow reducing the intake of sugars and provide bioactive phenolics (flavanones). The juices developed were ingested by volunteers ($n = 20$) and the resulting flavanones and their phase II metabolites in urine were analyzed by Ultra-High Performance Liquid Chromatography ElectroSpray Ionization Mass Spectrometry (UHPLC-ESI-MS/MS). A total of 16 metabolites were detected: eriodyctiol, naringenin, and homoeriodyctiol derivatives, where peak concentrations were attained 3.5 h after beverage intake. Sucralose and stevia were the sweeteners that provided the highest urinary excretion for most compounds. Sucrose did not provide a remarkable higher elimination through urine of any compounds in comparison with sucralose or stevia. These results propose two alternative sweeteners to sucrose (sucralose and stevia), an overused, high caloric sweetener that promotes some metabolic diseases.

Keywords: dietary intervention; maqui-citrus juice; flavanones; urinary excretion; UHPLC-ESI-QqQ-MS/MS

---

1. Introduction

Recently, scientific evidence has shown that sugar intake stimulates brain reward pathways [1–4], the basis for the consumers choosing the types of foods and beverages that use sweeteners such as sucrose, glucose, and saccharine liquid, among others. This consumption pattern has been related with the rising incidence of metabolic disorders, such as type 2 diabetes, by promoting obesity and insulin resistance [5–7]. Associated to these pathophysiological processes, an increase of strokes has also been reported [8]. Thus, in this aspect, sweetened beverages constitute the main dietary source of sugars in the human diet [9].

Global strategies against the impact on health of sugar consumption are focused on establishing balanced diets that avoid the excessive consumption of sweeteners [10]. In connection with this strategy, a number of studies have suggested the consumption of bioactive molecules (phytochemical compounds, such as flavanones) which have a vasodilator activity that contributes with the improvement of endothelial function, and thus, can help prevent the deleterious effects associated with sugar [10]. However, the bioavailability of these bioactive compounds is generally low (up to 10% of the total intake) and could be strongly conditioned by the physico-chemical properties of the matrix (foods and beverages). Therefore, a recent study has proposed the use of alternative sweeteners such as stevia, as a trans-glycosylated food additive that could potentially improve the stability, bioaccessibility, and bioavailability of polyphenols [11]. In addition, the strategy of using other sweeteners is in agreement with the recommendations by the World Health Organization (WHO) to reduce the intake of free sugars to values less than 10% in order to diminish the array of pathophysiological disorders associated with their consumption [12].

To contribute to the enhanced consumption of dietary sources of bioactive phytochemicals, in the last few years diverse fruits have been selected and characterized for the development of new bioactive beverages, which have a high bioavailability and content of bioactive compounds, and contribute with lowering the risk of a range of diseases [13–15]. In the work developed by Gironés-Vilaplana et al. [13], Maqui (*Aristotelia chilensis* (Mol.) Stuntz), a purple blackberry from Chile and Argentina, was chosen due to its importance as a "superfruit" according to its health properties, such as a high antioxidant capacity, cardioprotection activities, and inhibition of adipogenesis, and diabetes symptoms [13].

In addition to red fruits, citrus fruits have been selected as ingredients of the new beverage due to their phenolic composition, as they are flavonoid-rich fruits. Citrus juices have significant effects for decreasing diastolic blood pressure, enhancing endothelium-dependent microvascular reactivity and increasing the pro-coagulant activity [16,17]. On the other hand, this type of beverage contains many bioactive nutrients and non-nutrients, which can provide health benefits beyond nutrition and cardiovascular disease, cancer, diabetes, and obesity [18]. The most abundant polyphenols in citrus are flavanones, with naringenin (N), eriodyctiol (E), and hesperidin (H) being the most representative compounds of this family, with the latter being promoted as a preventive molecule against cardiovascular diseases [16].

This article deals with the influence of diverse sweeteners on the urinary excretion of flavanones on healthy humans after an acute administration of a polyphenols-rich beverage composed of maqui berry and citrus, and created using three different sweeteners, including sucrose (natural and high caloric), stevia (natural and non-caloric), and sucralose (artificial and non-caloric). These sweeteners were selected to compare a classical, natural, and high-caloric sweetener and two non-caloric alternatives. Stevia was selected as a natural and emergent sweetener and sucralose as an artificial and widely used sweetener.

2. Material and Methods

2.1. Chemicals and Reagents

The standards used for quantification purposes, eriodyctiol (E), homoeriodictyol (HE), naringenin (N), and hesperidin (H) were purchased from TransMIT (Geiben, Germany). Formic acid and acetonitrile were obtained from Fisher-Scientific (Loughborough, UK). All solutions were prepared with ultrapure water from a Milli-Q Advantage A10 ultrapure water purification system (Millipore, Burlington, MA, USA).

2.2. Juice Preparation and Characterization of the Phenolic Content

Fresh dry organic maqui powder was provided by Maqui New Life S.A. (Santiago, Chile). Cítricos de Murcia S.L. (Ceutí, Spain) and AMC Grupo Alimentación Fresco y Zumos S.A. (Espinardo, Spain) provided the citrus juices. Sucrose was provided by AB Azucarera Iberia S.L. (Madrid, Spain), Stevia by AgriStevia S.L. (Murcia, Spain), and Sucralose by Zukan (Murcia, Spain).

For the manufacturing of maqui-citrus juices, maqui powder was mixed with citrus juices to obtain the base beverage. Then, the three selected sweeteners were added to obtain the different beverages characterized in the present work. The beverages underwent a pasteurization treatment by heating them at 85 °C for 58 s. Afterwards, the mixtures were bottled and stored at 5 °C until consumption by the volunteers.

The polyphenolic composition of the beverages was also characterized. With this objective, the juices were centrifuged at 10500 rpm for 5 min (Sigma 1–13, B. Braun Biotech International, Osterode, Germany). The supernatants were filtered through a 0.45 mm polyvinylidene fluoride (PVDF) filter (Millex HV13, Millipore, Bedford, MA, USA) and analyzed by RP-HPLC-DAD. The identification and quantification of flavanones was performed by applying a previously-used method [13,19]. Briefly, chromatographic analyses of the samples for the identification and quantification of flavanones were carried out in a Luna 5 µm C18(2)100 Å column (250 mm × 4.6 mm), using Security Guard Cartridges PFD 4 mm × 3.0 mm both supplied by Phenomenex (Torrence, CA, USA). The solvents used for the chromatographic separation were Milli-Q water/formic acid (95.0:5.0, *v/v*) (solvent A) and methanol (solvent B), with a linear gradient (time (min.), %B) (0, 15%); (20, 30%); (30, 40%); (35, 60%); (40, 90%); (44, 90%); (45, 15%); and (50, 15%), using an Agilent Technologies 1220 Infinity Liquid Chromatograph, equipped with an auto-injector (G1313, Agilent Technologies) and a Diode Array Detector (1260, Agilent Technologies, Santa Clara, CA, USA). Chromatograms were recorded and processed on an Agilent ChemStation for LC 3D systems. The volume of injection and flow rate were 10 µL and 0.9 mL/min, respectively. The quantification of flavanones was done on UV chromatograms recorded as hesperidin at 280 nm and expressed as mg per 100 mL of juice.

2.3. Experimental Design

A double-blind, randomized, cross-sectional clinical study was conducted on overweight people (n = 20). The study and protocol were approved by the official Ethical Committee of Clinical Studies (CEIC) of the General University Hospital Morales Meseguer (Murcia), and registered at ClinicalTrials.gov (NCT04016337). The volunteers provided written consent to participate in this study. The criteria for the volunteers' selection for the study were to be in good health, overweight (between 24.9 and 29.9 kg/m^2 following WHO criteria), aged 40–60 years, non-smokers, non-dyslipidemic and normotense, with no chronic illnesses and not taking any medication. After an initial wash-out phase of 3 days with a strict diet free of polyphenols and added sugars, 330 mL of the test drinks (stevia, sucralose, and sucrose as the added sweetener) were administered on fasting conditions. Urine samples were collected 24 h prior (0 point), as well as in the following intervals: 0–3.5 h, 3.5–12.0 h, and 12.0–24.0 h. After 15 days, the process was repeated again, with the volunteers ingesting another drink developed with the remaining sweetener, until all the drinks were consumed by all the volunteers (3 rounds). The urine samples collected were stored at −80 °C until analysis. The analysis was performed once each period was finished and in the same batch to minimize analytical variations.

The total volume of each urine interval was recorded to calculate the absolute amounts of the compounds and metabolites excreted in the study period. Also, creatinine content was determined to normalize the concentrations of metabolites in urine as µg compound/mg creatinine, to control for differences in urine volumes.

2.4. Urine Samples Collection, Processing, and Analysis by UHPLC-ESI-MS/MS

Urine samples were defrosted and diluted 1:2 (*v/v*) in MilliQ-water/formic acid (99.9:0.1, *v/v*) and centrifuged at 15,000× *g* for 10 min, at 5 °C (Sigma 1–16, B. Braun Biotech International, Osterode, Germany). Afterwards, supernatants were filtered through 0.45 µm PVDF filters (Millex HV13, Millipore, Bedford, MA, USA) and stored at −20 °C until analysis by Ultra-High Performance Liquid Chromatography ElectroSpray Ionization Mass Spectrometry (UHPLC-ESI-MS/MS).

The identification and quantification of flavanone metabolites was performed by applying the method previously reported by Medina et al. with some modifications [20]. The analysis of samples on the profile and concentration of anthocyanin metabolites was carried out on with an Ascentis Express F5 column (5 cm × 2.1 mm; 2.7 µm) (Sigma, Osterode, Germany). The solvents used for the chromatographic separation were Milli-Q water/formic acid (99.9:0.1, *v/v*) (solvent A) and acetonitrile/formic acid (99.9:0.1, *v/v*) (solvent B), with a linear gradient (time (min.), %B) (0, 10%); (1, 10%); (10, 60%); (11, 80%); (13, 80%); (13.01, 10%), and (14.50, 10%); using an UHPLC system coupled with a triple quadrupole tandem mass spectrometer model 6460 (Agilent Technologies, Waldbronn, Germany), operating in multiple reaction monitoring (MRM) and negative/positive electrospray ionization (ESI) modes. The volume injected and flow rate were 10 µL and 0.2 mL/min, respectively. The MS parameters, at the optimized conditions, were gas temperature 325 °C; gas flow 10 L/min; nebulizer 40 psi; sheath gas heater 275 °C; sheath gas flow 12; capillary voltage 4000–5000 V; Vcharging 1000–2000. Data acquisition and processing were performed by using MassHunter software version B.08.00 (Agilent Technologies, Walbronn, Germany).

2.5. Statistical Analysis

Quantitative data are presented as mean ± SD of 20 volunteers. Specific differences were examined by an analysis of variance (ANOVA) and a multiple range test (Duncan's test). The data were processed using the SPSS 21.0 software package (SPSS Inc., Chicago, IL, USA.) and the level of significance was set at $p < 0.05$.

3. Results and Discussion

3.1. Flavanone Content of Juices

In order to establish the rate of urinary elimination of flavanones present in the maqui-citrus juices manufactured using three separate sweeteners (stevia, sucralose, and sucrose), their flavanones profile and concentration in juices were measured. In this regard, the presence of four flavanones was observed, which were found in the following decreasing concentration in all juices, H-rutinoside (4.87 mg/100 mL, on average) > N-rutinoside (1.31 mg/100 mL, on average) > E-rutinoside (0.32 mg/100 mL, on average) > N-hexoside derivatives (0.14 mg/100 mL, on average). As shown in Table 1, no significant differences were observed between the flavanone content of the maqui-citrus juices developed using diverse sweeteners either when considering individual or total flavanones.

Table 1. Flavanones composition of the maqui-citrus juices developed using diverse sweeteners.

Beverages	Flavanones [Z] (mg/100 mL)				
	N-Hexoside Derived	E-Rutinoside	N-Rutinoside	H-Rutinoside	Total
Stevia	0.15 ± 0.02	0.32 ± 0.04	1.30 ± 0.01	4.87 ± 0.01	6.64 ± 0.2
Sucralose	0.14 ± 0.02	0.32 ± 0.01	1.31 ± 0.01	4.86 ± 0.01	6.63 ± 0.1
Sucrose	0.14 ± 0.01	0.31 ± 0.03	1.31 ± 0.01	4.88 ± 0.01	6.64 ± 0.1
p-value	>0.05 [N.s.]	>0.05 [N.s.]	>0.05 [N.s.]	>0.05 [N.s.]	>0.05 [N.s.]

[Z] N, naringenin; E, eriodyctiol; H, hesperetin. N.s., sot significant.

3.2. Qualitative Analysis of Urine Metabolites of Flavanones from Maqui-Citrus Juice

To profile the metabolites of flavanones excreted in urine, 24-h urine collected after the ingestion of 330 mL of maqui-citrus juice by healthy volunteers were processed, allowing for the evaluation of the differences due to the sweetener employed in the development of the juices (stevia, sucralose, and sucrose). The analysis results show that the urine samples exhibited 16 diverse phenolic metabolites, derived from the list of flavanones present in the juices and shown in Table 2. More specifically, the compounds identified in the urine samples were E, E-glucuronide, E-sulfate, E-disulfate, HE, HE-glucuronide, HE-diglucuronide, HE-sulfate, HE-glucuronide-sulfate, N, N-glucoside, narirutin, N-glucuronide, N-diglucuronide, N-sulfate, and N-glucuronide-sulfate. Interestingly, neither H, nor its phase II derivative, were detected, although H was the main flavanone in the maqui-citrus based beverages, accounting for 73.3% of the total flavanones, on average.

Table 2. Qualitative analysis of flavanone metabolites in urine after the ingestion of maqui-citrus juices.

Compound	R.T. (min)	Precursor Ion	Product Ion	Fragmentation (V)	C.E. (V)	Polarity
Eriodyctiol metabolites						
Eriodyctiol (E)	6.49	287.0	151.0	70	10	Negative
Eriocitrin	N.f.	449.0	287.0	70	10	Negative
E-glucuronide	4.87	463.0	287.0	70	10	Negative
E-di-glucuronide	N.f.	639.0	287.0	70	10	Negative
E-sulfate	5.53	367.0	287.0	70	10	Negative
E-di-sulfate	4.24	447.0	287.0	70	10	Negative
E-glucuronide-sulfate	N.f.	543.0	287.0	70	10	Negative
Hesperetine metabolites						
Hesperetine (H)	7.30	302.0	151.0	70	20	Negative
Hesperidin	N.f.	609.0	302.0	70	20	Negative
H-glucuronide	N.f.	478.0	302.0	70	20	Negative
H-di-glucuronide	N.f.	664.0	302.0	70	20	Negative
H-sulfate	N.f.	382.0	302.0	70	20	Negative
H-di-sulfate	N.f.	462.0	302.0	70	20	Negative
H-glucuronide-sulfate	N.f.	558.0	302.0	70	20	Negative
Homoeriodyctiol metabolites						
Homoeriodyctiol (HE)	7.30	301.0	151.0	110	15	Negative
HE-glucuronide	5.50	477.0	301.0	110	15	Negative
HE-di-glucuronide	4.22	653.0	301.0	110	15	Negative
HE-sulfate	5.90	381.0	301.0	110	15	Negative
HE-di-sulfate	N.f.	461.0	301.0	110	15	Negative
HE-glucuronide-sulfate	4.67	557.0	301.0	110	15	Negative
Naringenin (N)	7.26	271.0	119.0	130	20	Negative
N-glucoside	4.63	433.0	271.0	130	20	Negative
Narirutin	4.86	579.0	271.0	130	20	Negative
N-glucuronide	5.07	433.0	271.0	130	20	Negative
N-di-glucuronide	4.09	623.0	271.0	130	20	Negative
N-sulfate	5.90	351.0	271.0	130	20	Negative
N-di-sulfate	N.f.	431.0	271.0	130	20	Negative
N-glucuronide-sulfate	4.87	527.0	271.0	130	20	Negative

C.E., collision Energy; N.f.—not found; R.T., retention time.

Since the metabolism of the precursors of flavanone metabolites is closely dependent on the metabolic traits of the volunteers and the inter-individual variation [21], a range of metabolites (E, E-disulfate, HE, HE-diglucuronide, and HE-glucuronide-sulfate) was found in urine of a reduced number of volunteers. These compounds were present in quantifiable amount but the limited number of volunteers excreting these molecules indicates that these were not representative. The inter-individual variation could also be responsible for the dispersion of the concentration of the metabolites, making difficult the identification of significant differences relative to the quantitative determinations. On the contrary, E-glucuronide, E-sulfate, HE-glucuronide, HE-sulfate, N, N-glucoside, narirutin, N-glucuronide, N-diglucuronide, N-sulfate, and N-glucuronide-sulfate were identified and quantified in the urine from all volunteers.

3.3. Quantification of Flavanone Metabolites in Urine Samples

In urine, the quantification of the flavanone metabolites excreted was based on basal urine (0 h), as well as on urine excreted between 0 and 3.5 h, 3.5–12.0 h, and 12.0–24.0 h. The excretion kinetics for E and N matched, showing the highest concentration at 3.5 h after the intake of the beverages (Figure 1). For this reason, all the concentrations described below referred to 3.5 h after the ingestion. In the case of HE derivatives, no significant increases in the urine concentration was found.

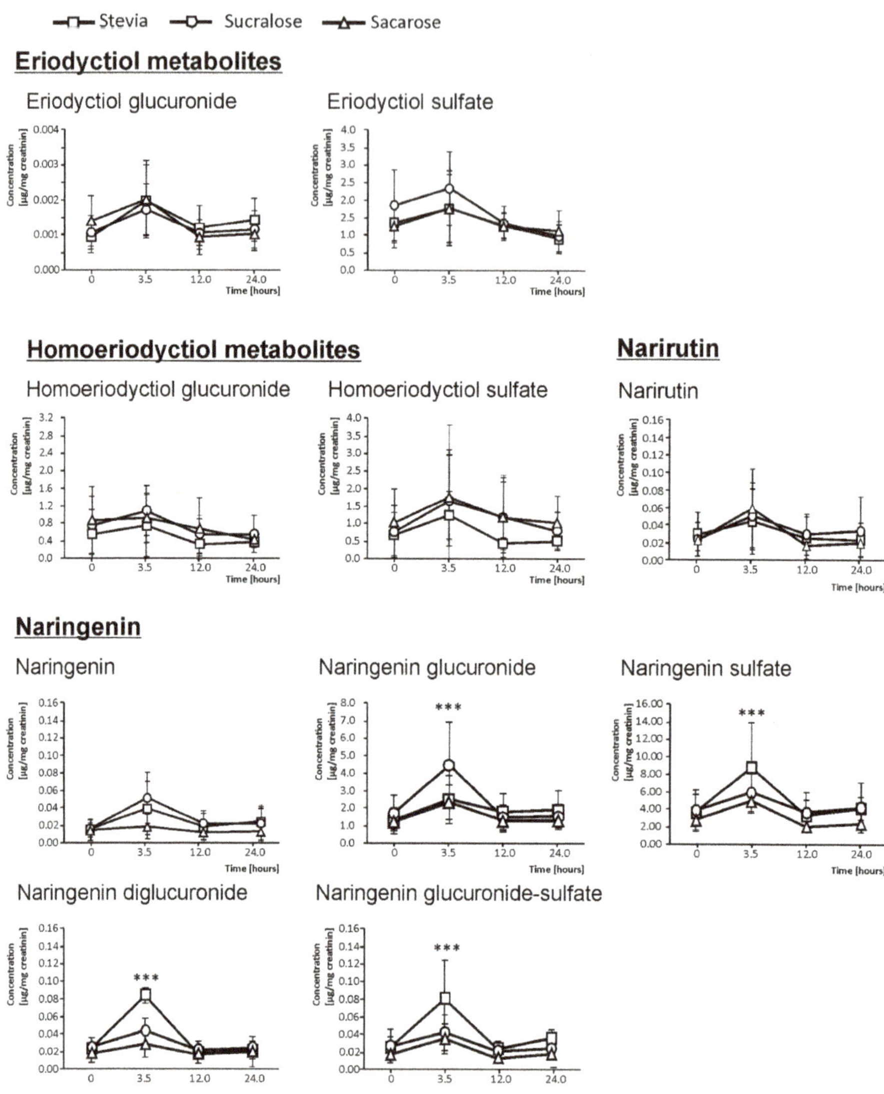

Figure 1. Content (mean ± SD, *n* = 2) of single flavanone metabolites in basal urine and 3.5, 12, and 24-h urine of healthy volunteers after ingesting 330 mL of maqui-citrus juices developed using as sweeteners stevia (□), sucralose (O), and sucrose (Δ). Significantly different bioavailabilities according to an analysis of variance (ANOVA) and Duncan's multiple rank test were found at $p < 0.001$ (***).

The sum of E and their phase II conjugates excretion provided values of 0.18 µg/mg creatinine in volunteers ingesting the beverages developed using sucralose as the sweetener. This concentration was higher than the one reached when using stevia and sucrose. The beverages developed using the latter sweeteners showed a 9.9% lesser amount of excretion than sucralose, on average. However, despite the trend observed, the large variation between volunteers did not allow for providing significant differences [21].

As mentioned above, the other flavanone metabolites present in urine in quantifiable concentrations were the N derivatives (Figure 1). The sum of N and their phase II derivatives excretion values was

115.91 µg/mg creatinine for sucralose, resulting in 12.2% and 32.1% lower amounts than sucralose, respectively, of stevia and sucrose-sweetened juices. When analyzing the effect of the sweetener in regard to the individual compounds, for N-diglucuronide, N-glucuronide-sulfate, and N-sulfate, the highest value corresponded to juices developed using stevia as a sweetener, which gave rise to 1.45, 1.69, and 19.74 µg/mg creatinine concentrations, respectively. Indeed, these concentrations significantly surpassed the urinary excretion reached when juices sweetened with sucralose and sucrose were ingested, by 33.3% (N-diglucuronide), 38.9% (N-glucuronide-sulfate), and 25.3% (N-sulfate), on average. In respect to N-glucuronide, the highest value corresponded to sucralose (93.71 µg/mg creatinine), which significantly surpassed the urine concentration provided by stevia and sucrose (25.2% lower, on average).

Moreover, stevia was the sweetener that provided a higher urinary excretion for most compounds derived from N (N-diglucuronide, N-glucuronide-sulfate, and N-sulfate), followed by sucralose (N-glucuronide). Sucrose did not provide remarkable higher rates of elimination through urine of any compounds in comparison with sucralose or stevia, as intestinal sugar carriers may play an important role in flavonoid absorption [22].

The results suggested that both stevia and sucralose were better than sucrose in terms of urinary excretion. Several studies of the effects on human health and metabolic diseases of stevia and sucralose showed contradictory results, as extensively reviewed by Daher et al. This author indicated that most intervention studies have assessed the role of isolated non-nutritional sweeteners and not as part of a habitual diet [23]. Thus, further studies are needed to learn more about the influence of stevia and sucralose on human health.

4. Conclusions

The results of the present work evidence that processing in respect to the selection of sweeteners does not seem to have any effect on flavanone concentration. On the other hand, the absorption rate of flavanones from citrus, excluding H, as they pass through the digestion system, is achieved through the formation of a variety of phase II derivatives. The results obtained pointed out sucralose and stevia as the sweeteners that had the greatest urinary excretion of N and most of its metabolites, which constitutes, as far as demonstrated responsible for valuable biological activities, a very useful marker for establishing the actual biological and healthy potential of the juices developed. Actually, they significantly surpassed the urinary excretion provided by sucrose that could be related with a variety of impacts of the diverse sweeteners on the actual bioavailability of flavanones. However, sucrose did not provide a higher urinary excretion as compared to sucralose or stevia in any of the cases.

This information would allow designing further studies of dietary interventions aimed at evaluating such sweeteners, which have gained relevance for human health. Indeed, considering the differences on urinary excretion between sweeteners, this study proposes two non-caloric sweetener alternatives (sucralose and stevia) in order to reduce the consumption of sucrose, a high caloric sweetener with an evident influence on metabolic disorders (type 2 diabetes and obesity, among others). Nevertheless, more studies are needed in order to better understand the effects on health of the two alternatives.

Author Contributions: Sampling, analytical determination, data processing, and drafting, V.A.; Experimental design, drafting, and supervision of the final version of the manuscript, R.D.-P.; Data analysis and revision of the final version of the manuscript, D.A.M.; Selection and management of volunteers and determination of urine creatinine, P.Z.; Supervised the project, designed the beverage formula, and contributed to the final version of the manuscript upon critical revision of the texts, C.G.-V. All authors have read and agreed to the published version of the manuscript.

Acknowledgments: This work was partially funded by the Spanish MINECO through Research Project AGL2016-75332-C2-1-R. VA was supported by a FPI grant (BES-2017-079754). The authors thank the English expert reviewer (Mario Fon) for the revision of the English style and grammar.

Conflicts of Interest: The authors declare no conflict of interest.

References

1. Avena, N.M.; Rada, P.; Hoebel, B.G. Evidence for sugar addiction: Behavioral and neurochemical effects of intermittent, excessive sugar intake. *Neurosci. Biobehav. Rev.* **2008**, *32*, 20–39. [CrossRef] [PubMed]
2. Hone-Blanchet, A.; Fecteau, S. Overlap of food addiction and substance use disorders definitions: Analysis of animal and human studies. *Neuropharmacology* **2014**, *85*, 81–90. [CrossRef] [PubMed]
3. Markus, C.R.; Rogers, P.J.; Brouns, F.; Schepers, R. Eating dependence and weight gain; no human evidence for a 'sugar-addiction' model of overweight. *Appetite* **2017**, *114*, 64–72. [CrossRef] [PubMed]
4. Smith, D.G.; Robbins, T.W. The neurobiological underpinnings of obesity and binge eating: A rationale for adopting the food addiction model. *Biol. Psychiatry* **2013**, *73*, 804–810. [CrossRef]
5. Malik, V.S.; Schulze, M.B.; Hu, F.B. Intake of sugar-sweetened beverages and weight gain: A systematic review. *Am. J. Clin. Nutr.* **2006**, *84*, 274–288. [CrossRef]
6. Palmer, J.R.; Boggs, D.A.; Krishnan, S.; Hu, F.B.; Singer, M.; Rosenberg, L. Sugar-Sweetened Beverages and Incidence of Type 2 Diabetes Mellitus in African American Women. *Arch. Intern. Med.* **2008**, *168*, 1487–1492. [CrossRef]
7. Schulze, M.B.; Manson, J.E.; Ludwig, D.S.; Colditz, G.A.; Stampfer, M.J.; Willett, W.C.; Hu, F.B. Sugar-Sweetened Beverages, Weight Gain, and Incidence of Type 2 Diabetes in Young and Middle-Aged Women. *JAMA* **2004**, *292*, 927–934. [CrossRef]
8. Bernstein, A.M.; de Koning, L.; Flint, A.J.; Rexrode, K.M.; Willett, W.C. Soda consumption and the risk of stroke in men and women. *Am. J. Clin. Nutr.* **2012**, *95*, 1190–1199. [CrossRef]
9. Reedy, J.; Krebs-Smith, S.M. Dietary sources of energy, solid fats, and added sugars among children and adolescents in the United States. *J. Am. Diet. Assoc.* **2010**, *110*, 1477–1484. [CrossRef]
10. Amiot, M.J.; Riva, C.; Vinet, A. Effects of dietary polyphenols on metabolic syndrome features in humans: A systematic review. *Obes. Rev.* **2016**, *17*, 573–586. [CrossRef]
11. Chen, L.; Cao, H.; Xiao, J. Polyphenols. In *Polyphenols: Properties, Recovery, and Applications*; Woodhead Publishing: Duxford, UK, 2018; pp. 45–67. [CrossRef]
12. Breda, J.; Jewell, J.; Keller, A. The Importance of the World Health Organization Sugar Guidelines for Dental Health and Obesity Prevention. *Caries Res.* **2019**, *53*, 149–152. [CrossRef] [PubMed]
13. Girones-Vilaplana, A.; Mena, P.; Moreno, D.A.; Garcia-Viguera, C. Evaluation of sensorial, phytochemical and biological properties of new isotonic beverages enriched with lemon and berries during shelf life. *J. Sci. Food Agric.* **2014**, *94*, 1090–1100. [CrossRef] [PubMed]
14. Sloan, A.E. Top 10 Functional Food Trends. *Food Technol.* **2018**, *72*, 26–43.
15. Törrönen, R.; McDougall, G.J.; Dobson, G.; Stewart, D.; Hellström, J.; Mattila, P.; Pihlava, J.-M.; Koskela, A.; Karjalainen, R. Fortification of blackcurrant juice with crowberry: Impact on polyphenol composition, urinary phenolic metabolites, and postprandial glycemic response in healthy subjects. *J. Funct. Foods* **2012**, *4*, 746–756. [CrossRef]
16. Morand, C.; Dubray, C.; Milenkovic, D.; Lioger, D.; Martin, J.F.; Scalbert, A.; Mazur, A. Hesperidin contributes to the vascular protective effects of orange juice: A randomized crossover study in healthy volunteers. *Am. J. Clin. Nutr.* **2011**, *93*, 73–80. [CrossRef] [PubMed]
17. Napoleone, E.; Cutrone, A.; Zurlo, F.; Di Castelnuovo, A.; D'Imperio, M.; Giordano, L.; De Curtis, A.; Iacoviello, L.; Rotilio, D.; Cerletti, C.; et al. Both red and blond orange juice intake decreases the procoagulant activity of whole blood in healthy volunteers. *Thromb. Res.* **2013**, *132*, 288–292. [CrossRef] [PubMed]
18. Gonzalez-Molina, E.; Dominguez-Perles, R.; Moreno, D.A.; Garcia-Viguera, C. Natural bioactive compounds of Citrus limon for food and health. *J. Pharm. Biomed. Anal.* **2010**, *51*, 327–345. [CrossRef]
19. Gironés-Vilaplana, A.; Mena, P.; García-Viguera, C.; Moreno, D.A. A novel beverage rich in antioxidant phenolics: Maqui berry (*Aristotelia chilensis*) and lemon juice. *LWT-Food Sci. Technol.* **2012**, *47*, 279–286. [CrossRef]
20. Medina, S.; Dominguez-Perles, R.; Garcia-Viguera, C.; Cejuela-Anta, R.; Martinez-Sanz, J.M.; Ferreres, F.; Gil-Izquierdo, A. Physical activity increases the urinary excretion of flavanones after dietary aronia-citrus juice intake in triathletes. *Food Chem.* **2012**, *135*, 2133–2137. [CrossRef]
21. Réveillon, T.; Rota, T.; Chauvet, É.; Lecerf, A.; Sentis, A. Repeatable inter-individual variation in the thermal sensitivity of metabolic rate. *Oikos* **2019**, *128*. [CrossRef]

22. Hollman, P.C.; de Vries, J.H.; van Leeuwen, S.D.; Mengelers, M.J.; Katan, M.B. Absorption of dietary quercetin glycosides and quercetin in healthy ileostomy volunteers. *Am. J. Clin. Nutr.* **1995**, *62*, 1276–1282. [CrossRef] [PubMed]
23. Daher, M.I.; Matta, J.M.; Abdel Nour, A.M. Non-nutritive sweeteners and type 2 diabetes: Should we ring the bell? *Diabetes Res. Clin. Pract.* **2019**, *155*, 107786. [CrossRef] [PubMed]

Article

A Randomized, Double-Blind, Placebo Controlled Trial to Determine the Effectiveness a Polyphenolic Extract (*Hibiscus sabdariffa* and *Lippia citriodora*) in the Reduction of Body Fat Mass in Healthy Subjects

Javier Marhuenda [1,*], Silvia Perez [1], Desirée Victoria-Montesinos [1], María Salud Abellán [1], Nuria Caturla [2], Jonathan Jones [2] and Javier López-Román [1]

1   Faculty of Health Sciences, San Antonio Catholic University of Murcia (UCAM), 30107 Murcia, Spain; sperez2@ucam.edu (S.P.); dvictoria@ucam.edu (D.V.-M.); mabellan@ucam.edu (M.S.A.); jlroman@ucam.edu (J.L.-R.)
2   Monteloeder S.L., 03203 Elche, Alicante, Spain; nuriacaturla@monteloeder.com (N.C.); jonathanjones@monteloeder.com (J.J.)
*   Correspondence: jmarhuenda@ucam.edu

Received: 21 November 2019; Accepted: 27 December 2019; Published: 6 January 2020

Abstract: The location and quantity of body fat determine the health risks, limiting people with obesity. Recently, polyphenols have attracted the attention of the scientific community because of their potential use for the reduction of obesity. A proprietary formula comprised of a blend of *Lippia citriodora* and *Hibiscus sabdariffa* has been recognized for its high content of polyphenols, powerful antioxidant molecules that may prevent weight gain and could be helpful for the treatment of obesity, as proven previously by in vivo models. The aim of the present study is to determine if the supplementation with *Lippia citriodora* and *Hibiscus sabdariffa* is useful for the treatment of obesity and/or weight control in subjects without a controlled diet. The intake of the extract for 84 days reduced body weight, the body mass index, and the fat mass measured with both bioimpedance and densitometry. This decrease in fat mass was observed to a greater extent, being significant, in the fat mass of the trunk (chest and torso).

Keywords: obesity; *Hibiscus sabdariffa*; *Lippia citriodora*; polyphenols

---

1. Introduction

The etiopathogenesis of obesity relies on the ability of adipocytes to transform the excess of energy into triglycerides, operating as an energy store. However, the excess of fatty drops in the adipocytes leads to insulin resistance, due to inflammatory responses, which limits their capacity to store the excess of energy [1]. The location and quantity of body fat determine the health risks, limiting people with obesity. That fact has a great impact on several metabolic and chronic diseases, including heart disease, cancer, arthritis, obstructive sleep apnea, hypertension, hyperlipidemia, and type 2 diabetes associated with insulin resistance [2].

This chronic condition of inflammation has especially been related to the generation of insulin resistance. Likewise, the inflammatory response modifies the metabolism of the organism, favoring or suppressing some pathways, such as insulin signaling pathway. That fact, when uncontrolled, may develop into a general inflammatory response than can perpetuate the chronic disfunction of the metabolism observed in obesity [3]. Due to the interest in natural extracts over the last 20 years [4], industry is strongly committed to nutraceuticals, either as a treatment for the disease or as a preventive treatment [5].

Polyphenols are natural extracts that have been extensively studied over the last few years due to their antioxidant and anti-inflammatory capacity, besides their possible role in the prevention and management of several diseases such as cardiovascular diseases, hypertension, diabetes, cancer, or neurodegenerative diseases [6–8]. Recently, polyphenols have attracted the attention of the scientific community because of their potential use in the reduction of obesity [9]. The proposed mechanisms include inhibition of the differentiation of adipocytes [10], enhanced fatty acid oxidation [8,11], diminished fatty acid extraction [12], increased thermogenesis and energy expenditure [13], or inhibition of digestive enzymes [14]. However, despite the popularity of this topic, the scientific literature is mainly based on animal and in vitro studies.

The basis of the present work is due to the lack of scientific reports on humans besides the promising use of *Lippia citriodora* and *Hibiscus sabdariffa*, recognized for their high content of polyphenols, powerful antioxidant molecules that may prevent several disease factors such as hypertension, oxidative stress, dyslipidemia, lipid mobilization, or endothelial stiffness [6]. The present *Hibiscus sabdariffa* and *Lippia citriodora* (HS-LC) formula has been previously studied in various clinical studies, in order to help induce weight loss during a controlled diet program [15,16]. The aim of the present study is to determine if the supplementation with *Lippia citriodora* and *Hibiscus sabdariffa* is useful for the treatment of obesity and/or weight control in the absence of a controlled diet.

2. Material and Methods

The study consisted of a double-blind, randomized, placebo controlled clinical trial, with two parallel branches of study depending on the extract consumed (experimental or placebo) and single center (Figure 1). A total of 84 sedentary and healthy subjects of both sexes was included in the study after matching all the including criteria (age between 18 and 65 years, both sexes, body mass index (BMI) between 25 and 35 kg/m^2) and none of the exclusion criteria (illness, pharmacological treatment, toxicological habits, or allergies). After a full disclosure of the implications and restrictions of the protocol, subjects were required to sign the informed consent from. Finally, subjects followed their regular diets during the whole study and were monitored by food consumption diaries.

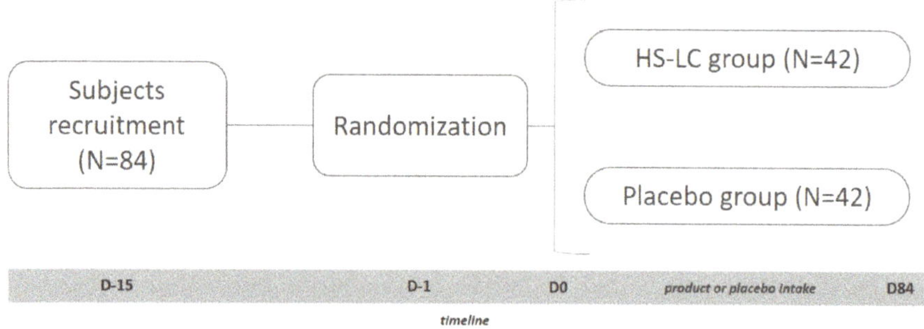

Figure 1. Graphic representation of the study. HS-LC, *Hibiscus sabdariffa* and *Lippia citriodora*

The extract under study (MetabolAid®) consisted of a capsule including a mixture of extracts from *Lippia citriodora* (LC) (325 mg) and *Hibiscus sabdariffa* (HS) (175 mg), both recognized for their abundance in sambubioside derivatives (HS) and verbascoside derivatives (LC) that may help to reduce problems associated with obesity [16]. The placebo capsules contained 500 mg of crystalline microcellulose, maintaining the same aspect as the product under study. MetabolAid® was provided by Monteloeder S.L. (Alicante, Spain) (Patent Application Number P201731147). The polyphenolic composition of the product was quantified and reported in previous studies [17,18].

The present study had a length of 84 days, in which subjects consumed the LC-HS extract or the placebo daily depending on the previous randomization. Therefore, 42 subjects were distributed in the placebo group, and the other 42 subjects were allocated to the LC-HS extract group. Each subject was assigned a code (generated by a number software generator (Epidat v4.1. Epidat, Spain) allocating them to one of the two study groups. Both the researchers and the subjects themselves did not know the composition of the groups.

Subjects came to the research center at the beginning of the study and at the end. At baseline, blood samples were obtained from the cubital veins of subjects from both groups. After blood collection and explaining the operation of the study, subjects received the LC-HS extract or placebo. In order to determine physical activity, every subject was equipped with an accelerometer (ActiGraph wGT3X-BT. ActiGraph, Pensacola, FL, USA) prior to the beginning of the study. Finally, the body composition of the volunteers (total fat mass, fat mass of the trunk, and fat mass of the lower body) was determined by two different methods, bioimpedance (Tanita BC-420M. Tanita Corporation, Arlington Heights, IL, USA) and densitometry (Norland XR-4 using Dual Photonic Absorciometry. Swissray, Edison, NJ, USA).

Moreover, physical activity could partly be a determinant of the possible observed fat loss during the study, due to the importance of exercise with respect to weight control. Therefore, in order to prove if some strategies are effective for the treatment of obesity, physical exercise must be measured and regulated. For this purpose, physical exercise was measured by the metabolic equivalent of task (MET), which led to the determination of metabolic equivalents, relating the intensity of physical activity to the kilocalories consumed by subjects. MET values were determined in accordance with the "Compendium of Physical Activities" [19].

Finally, blood analysis was performed for the monitoring of glycemic (glycemia and glycated hemoglobin) and lipidemic parameters (total cholesterol, low density lipoprotein (LDL), high density lipoprotein (HDL), and total triglycerides). The Reflotron Plus system (Roche Diagnostics S.L., Sant Cugat del Vallès, Barcelona, Spain) was used to obtain the glycemic and lipidemic values.

The study was conducted in accordance with the Declaration of Helsinki (randomized controlled trials registration number: NCT04105192), and the protocol was approved by the Ethics Committee of UCAM (CE011815).

3. Statistical Determinations

A descriptive analysis (mean and standard deviation) of all the variables under study was carried out. For quantitative variables, *t*-Student comparisons were developed between both the placebo and LC-HS extract groups. The qualitative variables were analyzed by means of a homogeneity test based on the Chi squared distribution when the expected values made it possible and by Fisher's exact test otherwise. To analyze the differences between the groups (LC-HS extract and placebo), an analysis of variance for repeated measures with the intrasubject factor (baseline and at the end, after 84 days) and intersubject factor (experimental group and placebo group) was carried out. The Bonferroni test was performed for the post-hoc analysis, using 0.05 as the level of significance. Statistical analysis was carried out with the SPSS 21.0 software (IBM. New York, NY, USA).

4. Results and Discussion

4.1. Blood Parameters

As shown in Table 1, blood parameters showed similar levels both at baseline and at the end of the study ($p > 0.05$).

Table 1. Evolution of biochemical blood parameters of subjects during the study.

		Baseline	Final
Total Cholesterol	Control	228.2 ± 6.2	232.8 ± 9.5
	EXTRACT	234.9 ± 6	223.2 ± 9
HDL	Control	56.5 ± 2.5	58.9 ± 2.4
	Extract	58.5 ± 2.1	60.6 ± 2.2
LDL	Control	144.1 ± 6.3	144.17.6
	Extract	134.1 ± 7.2	133.8 ± 7.5
Triglycerides	Control	138.8 ± 12.1	134.1 ± 612.4
	Extract	136 ± 612.4	133.1 ± 12.4

Values are expressed in mg/dL.

There was a downward trend for cholesterol and LDL values after the consumption of the product under study. However, the LC-HS extract did not improve ($p > 0.05$) the lipid profile of the subjects. In fact, the evolution of the lipid profile was similar between subjects consuming the placebo and those who consumed the LC-HS extract (Table 1). As observed for the lipid profile, the glycemic parameters showed similar values along all the study, regardless of the consumption of the placebo or the LC-HS extract (Table 1).

4.2. Weight and BMI

In order to determine the relevance of sex in the results obtained in the present research, we developed a sex based study of weight loss. Interestingly, men showed slightly high weight loss after treatment with LC-HS extract than women (90.03 ± 15.76 kg for men and 77.87 ± 77.05 kg for women at baseline compared to 89.02 ± 10.46 kg in the case of men and 77.05 ± 10.44 kg for women at the end of the study), but not for the placebo. However, neither weight nor BMI showed relevant changes, leading to no statistically significant variation ($p > 0.05$).

The weight of the subjects at the beginning of the study showed similar values for both the placebo and LC-HS extract. Placebo group subjects started the study with an 87.5 ± 15.8 kg body weight, while subjects who were included in the LC-HS extract group showed a total of 87.4 ± 11.1 kg body weight (Figure 2). When comparing the weight of the volunteers of both groups under study, no statistically significant differences were observed ($p = 0.972$); hence, it can be stated that the weight of all the subjects under study was similar at the beginning of the present study.

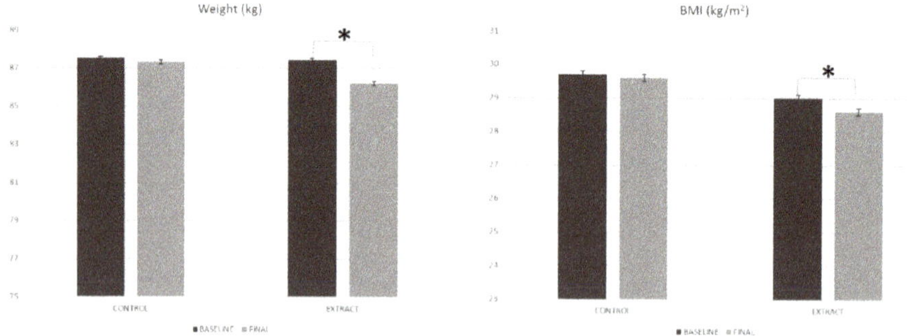

Figure 2. Weight and BMI evolution along the study. * Means statistically significant differences.

After 84 days of treatment, the weight of subjects from the placebo group was 87.3 ± 15.3 kg (Figure 2), which represents a similar body weight as that observed at baseline ($p = 0.479$). On the other hand, after 84 days of treatment with the experimental LC-HS extract, the body weight of subjects was

86.2 ± 10.9 kg. That change in the body weight of volunteers consuming the polyphenol rich LC-HS extract for 84 days represented a statistically significant variation ($p < 0.001$) compared to the body weight observed at baseline in these subjects. Therefore, it can be determined that the LC-HS extract significantly decreased the body weight of the subjects ($p = 0.014$).

Similar to what was observed previously for body weight, the BMI of the subjects at the beginning of the study showed similar values ($p = 0.436$) for the placebo group (29.7 ± 4.7 kg/m^2) and HS-LC group (29.0 ± 3.1 kg/m^2). After the consumption of both the placebo and the LC-HS extract for 84 days, the BMI was 29.6 ± 4.4 kg/m^2 and 28.6 ± 3.0 kg/m^2 for the placebo and LC-HS extract, respectively. Therefore, the diminution of BMI after the consumption of the placebo was shown to be less than that observed for the LC-HS extract ($p = 0.013$).

Despite the lack of in vivo studies found in the scientific literature, the antiobesogenic properties of LC-HS has been previously reported in some in vitro models and animal models. For one study using an insulin resistant hypertrophic 3T3-L1-adipocyte model, the polyphenols from HS were shown to inhibit the increase in triglycerides, oxidative stress due to free radicals, and inflammatory adipokines [18,20]. In addition, the administration of HS in the form of an extract has also been able to prevent hepatic steatosis by regulating the expression of various genes responsible for glycemic and lipid homeostasis [21], as well as reducing blood pressure and improving vascular endothelial function [11] in hyperlipidemic mouse models [22]. Finally, despite HS being a rich polyphenolic matrix, the scientific literature seems to point to quercetin-3-O-β-D-glucuronide and its aglycone as the main ones responsible for the effects described [23].

In the same way, other studies of LC and its content in polyphenols showed promising effects in the reduction of fatty tissue, such as decreased lipogenesis, enhanced oxidation of fatty acids, and the activation of the AMP-activated protein kinase (AMPK) pathway (probably through the activation of the peroxisome proliferator-activated receptor (PPAR-gamma) receptor and adiponectin) [24]. In the same way as the HS extract, the treatment with LC in an animal model with hyperlipidemia prevented fatty liver disease and improved lipid metabolism. In fact, it seems that LC-HS can have synergistic effects for the treatment of fatty liver and lipid metabolism [22,24]. A recent study has revealed the joint ability of these two plant matrices to treat obesity, improving the metabolism of hyperlipidemic rats through the increase of thermogenesis inducing genes in white adipose tissue and correlating with the increase in AMPK phosphorylation and fatty acid oxidation at the liver level [16]. Moreover, in one of the few in vivo studies with humans, it was reported that the combination of LC-HS can modulate the response of incretins, related to the regulation of appetite and hunger, which could be beneficial in the treatment of obesity in the context of an isocaloric healthy diet [15].

Finally, Herranz-López and Barrajón-Catalán [24] studied the effects of LC in an insulin resistant hypertrophic 3T3-L1-adipocyte model and reported a decrease in lipogenesis, improvement of fatty acid oxidation, and activation of the energy regulator pathway regulated by AMPK. They suggested a mechanism activated by transcriptional factors such as reactive oxygen species (ROS) mediated downregulation of nuclear factor kappa-B transcription factor (NF-κB) and peroxisome proliferator activated receptor gamma (PPAR-γ) dependent transcriptional upregulation of adiponectin [25].

4.3. Body Composition

4.3.1. Bioimpedance

Weight reduction is necessary for the treatment of obesity. However, some treatments may cause a reduction on the muscular mass, reducing metabolic waste, and compromising weight loss and/or maintenance of the lost weight [26]. In order to guarantee that the loss of body weight is accompanied by the loss of fat mass and the maintenance of muscular mass, we performed two different analyses of body composition.

Bioimpedance is a secure, easy, and non-invasive method that can be used for the measurement of body composition [27]. However, the values must be considered carefully due to the limited reliability of bioimpedance. Intracellular water alterations are frequent in protein-calorie malnutrition, and therefore, the non-fatty mass measures do not exactly reflect the amount of real body composition [27]. Ellis and Bell [28] described a series of general recommendations for the use of impedance after its implementation became widespread by a large number of researchers and has not always been well used. In general, the impedance of legs and arms is less predictive of non-fatty mass than full-body bioimpedance [29].

After the 84 days of treatment, the values obtained in the present study (Figure 3) showed a reduction in the fat mass in the case of the LC-HS extract group ($p < 0.001$), but no differences were observed for the placebo group ($p = 0.309$). Differences between both groups were also statistically significant ($p < 0.002$), exhibiting a remarkable capacity of the LC-HS extract to reduce body fat mass. Moreover, it was complemented by the maintenance of muscle mass, as observed in the Figure 3.

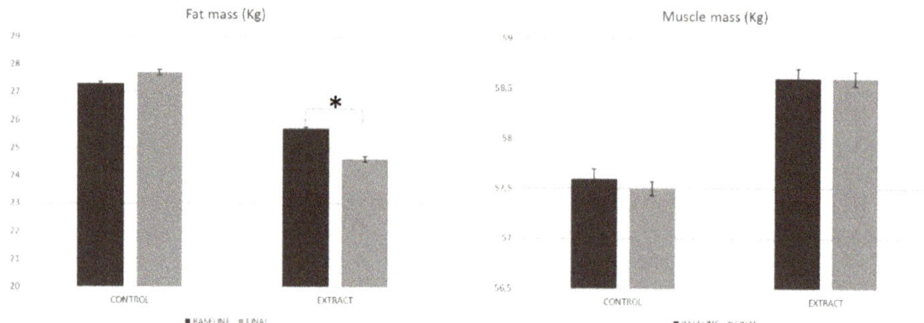

Figure 3. Body composition evolution along the study, measured by bioimpedance. * Means statistically significant differences.

4.3.2. Densitometry

Due to the limited reliability of bioimpedance, the analysis of body composition was repeated using densitometry equipment. Dual-energy X-ray absorptiometry (DXA) has been generally used for clinical purposes to measure bone mineral content and bone mineral density as part of osteoporosis evaluation [30]. More recently, densitometry has gained popularity for other purposes due to its ability to measure body composition, incorporating measures of whole body and regional lean mass and fat mass, including visceral adipose tissue [31,32].

The total fat mass of the subjects at the beginning of the study showed similar values for both the placebo and LC-HS extract groups. The placebo group subjects started the study with a 31.4 ± 1.8 kg fat mass, while subjects who were included in the extract group showed a total of 32.3 ± 1.7 kg fat mass (Figure 4). When comparing the fat mass of the volunteers of both groups under study, no statistically significant differences were observed ($p = 0.713$), so it can be stated that the fat mass of all the subjects under study was similar at the beginning of the present study.

At the end of the study, the fat mass of subjects from the placebo group was 31.0 ± 1.7 kg (Figure 4), which represents a similar fat mass as that observed at baseline ($p = 0.361$). In turn, the fat mass of subjects consuming the LC-HS extract was statistically less ($p > 0.001$) compared to that observed at baseline (30.1 ± 1.7 kg), which represents a 6.9% decrease with respect to baseline. Interestingly, the differences observed in the reduction of fat mass showed the effectiveness of the HS-LC extract in reducing fat mass.

Fat deposits observed in the obese population determine comorbidities and many related illnesses [33]. Central fat deposits have been particularly associated with alterations in various metabolic pathways more than peripheral fat, being most evident when intra-abdominal visceral fat deposits increase [34]. Visceral obesity has also been associated with endocrine abnormalities, especially with regard to cortisol, growth hormones, and sex steroids, with a profound impact on the activity of these hormones in peripheral or white tissues [33]. Individuals with visceral obesity and/or metabolic syndrome virtually present all hormonal alterations that occur in old age, suggesting that this condition determines a kind of premature aging [34].

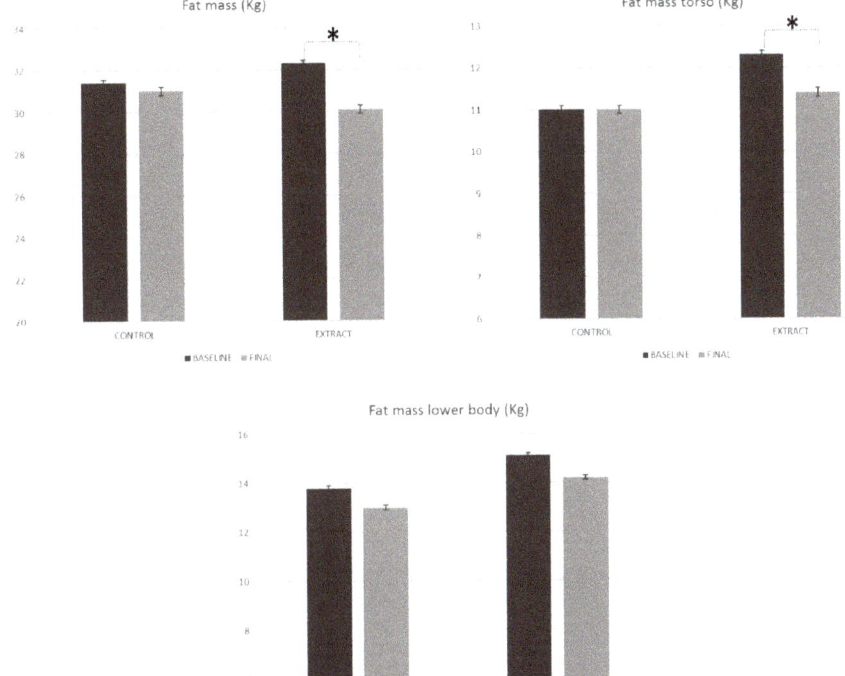

Figure 4. Body composition evolution along the study, measured by densitometry. * Means statistically significant differences.

In the present study, the placebo group did not reduce fat mass neither in the torso ($p = 0.913$) nor the lower body ($p = 0.391$). Regarding the intake of the LC-HS extract, the results from the densitometry analysis of the body weight showed a statistically significant diminution of the fat mass allocated in the torso ($p = 0.001$) and a minor reduction in the lower body ($p = 0.306$) (Figure 4).

Judging by the evolution of both groups and the difference observed ($p = 0.018$), the LC-HS extract was more effective at decreasing fat mass in the torso after 84 days of treatment, compared with placebo. However, despite reducing fat mass in the lower body, the results were not statistically significant for the placebo or the LC-HS extract. As observed, treatment with the LC-HS extract for 84 days, but not placebo, led to the reduction of central fat mass, which could reduce the comorbidities associated with obesity and the related inflammation.

Of all the proposed mechanisms, the regulation of AMPK expression appeared to be the main molecular target of the action of LC-HS, allowing glycidic and lipid regulation in the body [16,24]. Activation of AMPK increases the biogenesis and density of mitochondria, which would lead to an increase in the maximum and basal metabolic rate, as well as tolerance to physical activity [35]. In fact, the increase of AMPK expression in mice seems to be decisive to increase voluntary physical activity [36]. Finally, AMPK also improves the metabolic rate by activating phosphofructokinase-2, increasing gluconeogenesis, and reducing the sensation of appetite [37].

Therefore, the activation of AMPK by HS and LC could increase the basal metabolic rate, increasing energy expenditure. This would lead to an improvement in body composition, necessary for the treatment of obesity [38]. Finally, the activation of AMPK in certain regions of the hypothalamus responsible for the control of food intake decreases the feeling of appetite and increases the action of leptin [39].

4.4. Physical Activity

In order to reduce the error caused by personal differences in physical activity [40], sedentary volunteers were exclusively recruited. The measurement of METs by the accelerometer revealed that subjects from both placebo and experimental groups maintained the same physical activity during the 84 days of the study (Figure 5). Regarding placebo group, values ranged ($p = 0.418$) from 1.7 ± 0.3 MET at baseline to 1.8 ± 0.3 MET at the end of the study. In turn, the experimental group showed similar values ($p = 0.842$) both at baseline 1.7 ± 0.4 MET and at the end of the study 1.7 ± 0.3 MET. That fact supports that changes observed in subjects consuming the LC-HS extract may be a consequence of such consumption and not a change in physical activity habits.

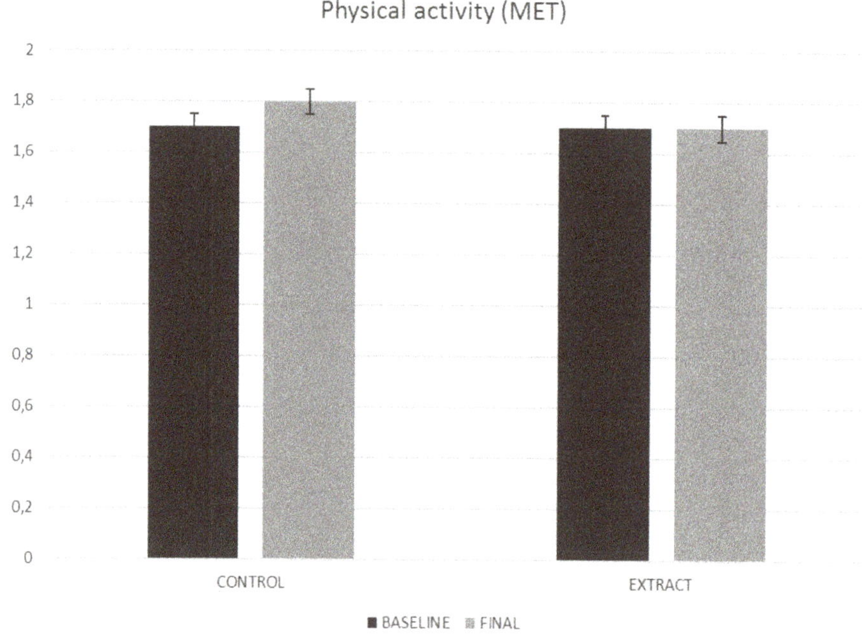

Figure 5. Physical activity EVOLUTION along the study. MET, metabolic equivalent of task.

5. Conclusions

The daily consumption of the LC-HS extract, but not the placebo, for 84 days was able to reduce body weight, BMI, and central fat mass, regardless of the physical activity. The LC-HS extract seems to affect only weight variables due to possible changes in molecular pathways, but its impact on biochemistry blood parameters such as cholesterol, LDL, HDL, and triglycerides needs more study.

Author Contributions: Funding acquisition: N.C., J.J. and J.L.-R.; investigation: J.M. and J.L.-R.; methodology: J.M., S.P., D.V.-M., M.S.A., and J.L.-R.; project administration: J.J.; resources: N.C.; supervision: J.M.; validation: J.L.-R.; writing, review and editing, J.M. and J.L.-R. All authors have read and agreed to the published version of the manuscript.

Funding: This research received external funding from the NNOPREFAT Project, GA 783838, in the framework of the Horizon 2020 Programme, "Natural Food Formulation for the Prevention and Treatment of Obesity and Metabolic Syndrome Obtained with Herbal Extracts".

Conflicts of Interest: The authors declare no conflict of interest.

References

1. Díaz, M. Presente y futuro del tratamiento farmacológico de la obesidad. *Present Future Pharmacol. Treat. Obes.* **2005**, *73*, 137–144.
2. Blancas-Flores, G.; Almanza-Pérez, J.C.; López-Roa, R.I.; Alarcón-Aguilar, F.J.; García-Macedo, R.; Cruz, M. La obesidad como un proceso inflamatorio. *Boletín médico del Hospital Infantil de México* **2010**, *67*, 88–97.
3. Grundy, S.M. *Metabolic Syndrome*; Springer: Cham, Switzerland, 2018.
4. Alvídrez-Morales, A.; González-Martínez, B.E.; Jiménez-Salas, Z. Tendencias en la producción de alimentos: Alimentos funcionales. Available online: https://www.medigraphic.com/pdfs/revsalpubnut/spn-2002/spn023g.pdf (accessed on 30 November 2019).
5. Anchia, I.A. *Alimentos y Nutrición en La Práctica Sanitaria*; Ediciones Díaz de Santos: Madrid, Spain, 2003.
6. Serban, C.; Sahebkar, A.; Ursoniu, S.; Andrica, F.; Banach, M. Effect of sour tea (*Hibiscus sabdariffa* L.) on arterial hypertension: A systematic review and meta-analysis of randomized controlled trials. *J. Hypertens.* **2015**, *33*, 1119–1127. [CrossRef] [PubMed]
7. Marhuenda, J.; Medina, S.; Martínez-Hernández, P.; Arina, S.; Zafrilla, P.; Mulero, J.; Oger, C.; Galano, J.-M.; Durand, T.; Ferreres, F.; et al. Melatonin and hydroxytyrosol protect against oxidative stress related to the central nervous system after the ingestion of three types of wine by healthy volunteers. *Food Funct.* **2017**, *8*, 64–74. [CrossRef]
8. Marhuenda, J.; Medina, S.; Martínez-Hernández, P.; Arina, S.; Zafrilla, P.; Mulero, J.; Oger, C.; Galano, J.-M.; Durand, T.; Solana, A. Effect of the dietary intake of melatonin-and hydroxytyrosol-rich wines by healthy female volunteers on the systemic lipidomic-related oxylipins. *Food Funct.* **2017**, *8*, 3745–3757. [CrossRef]
9. Zaki, M.E.; Hala, T.; El-Gammal, M.; Kamal, S. Indicators of the metabolic syndrome in obese adolescents. *Arch. Med. Sci. AMS* **2015**, *11*, 92. [CrossRef]
10. Min, S.Y.; Yang, H.; Seo, S.G.; Shin, S.H.; Chung, M.Y.; Kim, J.; Lee, S.J.; Lee, H.J.; Lee, K.W. Cocoa polyphenols suppress adipogenesis in vitro and obesity in vivo by targeting insulin receptor. *Int. J. Obes.* **2013**, *37*, 584. [CrossRef]
11. Shimoda, H.; Tanaka, J.; Kikuchi, M.; Fukuda, T.; Ito, H.; Hatano, T.; Yoshida, T. Effect of polyphenol-rich extract from walnut on diet-induced hypertriglyceridemia in mice via enhancement of fatty acid oxidation in the liver. *J. Agric. Food Chem.* **2009**, *57*, 1786–1792. [CrossRef]
12. Matsui, N.; Ito, R.; Nishimura, E.; Yoshikawa, M.; Kato, M.; Kamei, M.; Shibata, H.; Matsumoto, I.; Abe, K.; Hashizume, S. Ingested cocoa can prevent high-fat diet-induced obesity by regulating the expression of genes for fatty acid metabolism. *Nutrient* **2005**, *21*, 594–601. [CrossRef]
13. Stohs, S.J.; Badmaev, V. A review of natural stimulant and non-stimulant thermogenic agents. *Phytother. Res.* **2016**, *30*, 732–740. [CrossRef]
14. Gu, Y.; Hurst, W.J.; Stuart, D.A.; Lambert, J.D. Inhibition of key digestive enzymes by cocoa extracts and procyanidins. *J. Agric. Food Chem.* **2011**, *59*, 5305–5311. [CrossRef] [PubMed]
15. Boix-Castejón, M.; Herranz-López, M.; Gago, A.P.; Olivares-Vicente, M.; Caturla, N.; Roche, E.; Micol, V. Hibiscus and lemon verbena polyphenols modulate appetite-related biomarkers in overweight subjects: A randomized controlled trial. *Food Funct.* **2018**, *9*, 3173–3184. [CrossRef] [PubMed]
16. Lee, Y.S.; Yang, W.K.; Kim, H.; Min, B.; Caturla, N.; Jones, J.; Park, Y.-C.; Lee, Y.-C.; Kim, S.H. Metabolaid® combination of lemon verbena and hibiscus flower extract prevents high-fat diet-induced obesity through AMP-activated protein kinase activation. *Nutrient* **2018**, *10*, 1204. [CrossRef] [PubMed]
17. Ahmad-Qasem, M.H.; Caánovas, J.; Barrajón-Catalán, E.; Carreres, J.E.; Micol, V.; García-Pérez, J.V. Influence of olive leaf processing on the bioaccessibility of bioactive polyphenols. *J. Agric. Food Chem.* **2014**, *62*, 6190–6198. [CrossRef]
18. Herranz-López, M.; Fernández-Arroyo, S.; Pérez-Sanchez, A.; Barrajón-Catalán, E.; Beltrán-Debón, R.; Menéndez, J.A.; Alonso-Villaverde, C.; Segura-Carretero, A.; Joven, J.; Micol, V. Synergism of plant-derived polyphenols in adipogenesis: Perspectives and implications. *Phytomed* **2012**, *19*, 253–261.

19. Ainsworth, B.E.; Haskell, W.L.; Herrmann, S.D.; Meckes, N.; Bassett Jr, D.R.; Tudor-Locke, C.; Greer, J.L.; Vezina, J.; Whitt-Glover, M.C.; Leon, A.S. Compendium of physical activities: A second update of codes and MET values. *Med. Sci. Sports Exerc.* **2011**, *43*, 1575–1581. [CrossRef]

20. Beltrán-Debón, R.; Alonso-Villaverde, C.; Aragones, G.; Rodriguez-Medina, I.; Rull, A.; Micol, V.; Segura-Carretero, A.; Fernandez-Gutierrez, A.; Camps, J.; Joven, J. The aqueous extract of *Hibiscus sabdariffa* calices modulates the production of monocyte chemoattractant protein-1 in humans. *Phytomed* **2010**, *17*, 186–191. [CrossRef]

21. Joven, J.; March, I.; Espinel, E.; Fernández-Arroyo, S.; Rodríguez-Gallego, E.; Aragonès, G.; Beltran-Debon, R.; Alonso-Villaverde, C.; Rios, L. *Hibiscus sabdariffa* extract lowers blood pressure and improves endothelial function. *Mol. Nutr. Food Res.* **2014**, *58*, 1374–1378. [CrossRef]

22. Joven, J.; Espinel, E.; Rull, A.; Aragonès, G.; Rodríguez-Gallego, E.; Camps, J.; Micol, V.; Herranz-Lopez, M.; Menendez, J.A.; Borras, I.; et al. Plant-derived polyphenols regulate expression of miRNA paralogs miR-103/107 and miR-122 and prevent diet-induced fatty liver disease in hyperlipidemic mice. *Biochim. et Biophys. Acta (BBA)-Gen. Subj.* **2012**, *1820*, 894–899. [CrossRef]

23. Olivares-Vicente, M.; Barrajon-Catalan, E.; Herranz-Lopez, M.; Segura-Carretero, A.; Joven, J.; Encinar, J.A.; Micol, V. Plant-derived polyphenols in human health: Biological activity, metabolites and putative molecular targets. *Curr. Drug Metab.* **2018**, *19*, 351–369. [CrossRef]

24. Herranz-López, M.; Barrajón-Catalán, E.; Segura-Carretero, A.; Menéndez, J.A.; Joven, J.; Micol, V. Lemon verbena (*Lippia citriodora*) polyphenols alleviate obesity-related disturbances in hypertrophic adipocytes through AMPK-dependent mechanisms. *Phytomed* **2015**, *22*, 605–614.

25. Shirwany, N.A.; Zou, M.H. AMPK: A cellular metabolic and redox sensor: A mini review. *Front. Biosci.* **2014**, *19*, 447. [CrossRef] [PubMed]

26. Nuttall, F.Q. Body mass index: Obesity, BMI, and health: A critical review. *Nutr. Today* **2015**, *50*, 117. [CrossRef] [PubMed]

27. Barbosa-Silva, M.C.G.; Barros, A.J.; Post, C.L.; Waitzberg, D.L.; Heymsfield, S.B. Can bioelectrical impedance analysis identify malnutrition in preoperative nutrition assessment? *Nutrient* **2003**, *19*, 422–426. [CrossRef]

28. Ellis, K.J.; Bell, S.J.; Chertow, G.M.; Chumlea, W.C.; Knox, T.A.; Kotler, D.P.; Lukaski, H.C.; Schoeller, D.A. Bioelectrical impedance methods in clinical research: A follow-up to the NIH Technology Assessment Conference. *Nutrient* **1999**, *15*, 874–880. [CrossRef]

29. Alvero-Cruz, J.R.; Gómez, L.C.; Ronconi, M.; Vázquez, R.F.; i Manzañido, J.P. La bioimpedancia eléctrica como método de estimación de la composición corporal: Normas prácticas de utilización. *Rev. Andal. Med. Deporte* **2011**, *4*, 167–174.

30. Blake, G.M.; Fogelman, I. The clinical role of dual energy X-ray absorptiometry. *Mag. Andal. Med. Sport.* **2009**, *71*, 406–414. [CrossRef]

31. Meyer, N.L.; Sundgot-Borgen, J.; Lohman, T.G.; Ackland, T.R.; Stewart, A.D.; Maughan, R.J.; Smith, S.; Müller, W. Body composition for health and performance: A survey of body composition assessment practice carried out by the Ad Hoc Research Working Group on Body Composition, Health and Performance under the auspices of the IOC Medical Commission. *Br J Sports Med.* **2013**, *47*, 1044–1053. [CrossRef]

32. Ackland, T.R.; Lohman, T.G.; Sundgot-Borgen, J.; Maughan, R.J.; Meyer, N.L.; Stewart, A.D.; & Müller, W. Current status of body composition assessment in sport. *Sports Med.* **2012**, *42*, 227–249. [CrossRef]

33. Huang, T.T.K.; Johnson, M.S.; Figueroa-Colon, R.; Dwyer, J.H.; Goran, M.I. Growth of visceral fat, subcutaneous abdominal fat, and total body fat in children. *Obes. Res.* **2001**, *9*, 283–289. [CrossRef]

34. Gutiérrez, S.A.G.; Orozco, G.E.M.; Rodríguez, E.M.; Vázquez, J.D.J.S.; Camacho, R.B. La grasa visceral y su importancia en obesidad. *J. Endocrinol. Nutr.* **2002**, *10*, 121–127.

35. Ross, T.T.; Overton, J.D.; Houmard, K.F.; Kinsey, S.T. β-GPA treatment leads to elevated basal metabolic rate and enhanced hypoxic exercise tolerance in mice. *Physiol. Rep.* **2017**, *5*, e13192. [CrossRef] [PubMed]

36. Thomson, D.M.; Porter, B.B.; Tall, J.H.; Kim, H.J.; Barrow, J.R.; Winder, W.W. Skeletal muscle and heart LKB1 deficiency causes decreased voluntary running and reduced muscle mitochondrial marker enzyme expression in mice. *A. J. Physiol.-Endocrinol. Metab.* **2007**, *292*, E196–E202. [CrossRef] [PubMed]

37. Chaube, B.; Malvi, P.; Singh, S.V.; Mohammad, N.; Viollet, B.; Bhat, M.K. AMPK maintains energy homeostasis and survival in cancer cells via regulating p38/PGC-1α-mediated mitochondrial biogenesis. *Cell Death Discov.* **2015**, *1*, 15063. [CrossRef] [PubMed]

38. Dal-Pan, A.; Blanc, S.; Aujard, F. Resveratrol suppresses body mass gain in a seasonal non-human primate model of obesity. *BMC Physiol.* **2010**, *10*, 11. [CrossRef] [PubMed]
39. Weltman, A.; Levine, S.; Seip, R.L.; Tran, Z.V. Accurate assessment of body composition in obese females. *Am. J. Clin. Nutr.* **1988**, *48*, 1179–1183. [CrossRef]
40. Sanz, J.M.M.; Otegui, A.U.; Mielgo-Ayuso, J. Necesidades energéticas, hídricas y nutricionales en el deporte. *Eur. J. Hum. Mov.* **2013**, *30*, 37–52.

Correction

Correction: Marhuenda, J., et al. A Randomized, Double-Blind, Placebo Controlled Trial to Determine the Effectiveness a Polyphenolic Extract (*Hibiscus sabdariffa* and *Lippia citriodora*) in the Reduction of Body Fat Mass in Healthy Subjects. *Foods* 2020, 9(1), 55

Javier Marhuenda [1,*], Silvia Perez-Piñero [1], Desirée Victoria-Montesinos [1], María Salud Abellán-Ruiz [1], Nuria Caturla [2], Jonathan Jones [2] and Javier López-Román [1]

[1] Faculty of Health Sciences, San Antonio Catholic University of Murcia (UCAM), 30,107 Murcia, Spain; sperez2@ucam.edu (S.P.-P.); dvictoria@ucam.edu (D.V.-M.); mabellan@ucam.edu (M.S.A.-R.); jlroman@ucam.edu (J.L.-R.)

[2] Monteloeder S.L., 03203 Elche, Alicante, Spain; nuriacaturla@monteloeder.com (N.C.); jonathanjones@monteloeder.com (J.J.)

* Correspondence: jmarhuenda@ucam.edu

Received: 14 February 2020; Accepted: 25 February 2020; Published: 3 March 2020

The authors wish to make the following correction to this paper [1]:

There are text errors in Section 4.3.2 Densitometry, third paragraph. The second sentence states that "...LC-HS extract was statistically less ($p > 0.001$) compared...". It should instead read "...LC-HS extract was statistically less ($p < 0.001$) compared... ". Furthermore, the last sentence of the same paragraph reads "Interestingly, the differences observed in the reduction of fat mass showed the effectiveness of the HS-LC extract in reducing fat mass". It should instead read "Interestingly, the differences observed in the reduction of fat mass showed the effectiveness of the LC-HS extract in reducing fat mass". In this respect, there are several parts of the manuscript which reads HS-LC, it should be read as LC-HS instead. This includes: page 2, paragraph 2; Figure 1; page 4, last paragraph.

There is also a missing explanation in the same section, regarding the intergroup analysis of the results. At the end of the third paragraph, it should include the following sentence: "Intergroup analysis revealed significant differences between both groups, where LC-HS significantly decreased body fat mass compared to placebo ($p < 0.003$)".

In Section 4.2 Weight and BMI, the sentence "Therefore, it can be determined that the LC-HS extract significantly decreased the body weight of the subjects ($p = 0.014$)", should read "Inter-group analysis revealed that the LC-HS extract significantly decreased the body weight compared to placebo ($p = 0.014$)". In the same section, sentence "Therefore, the diminution of BMI after the consumption of the placebo was shown to be less than that observed for the LC-HS extract ($p = 0.013$)", should read "Therefore, intergroup analysis revealed that the BMI of the LC-HS group significantly decreased with respect to the placebo group ($p = 0.013$)".

Information is missing from Figure 2. The original figure is as follows:

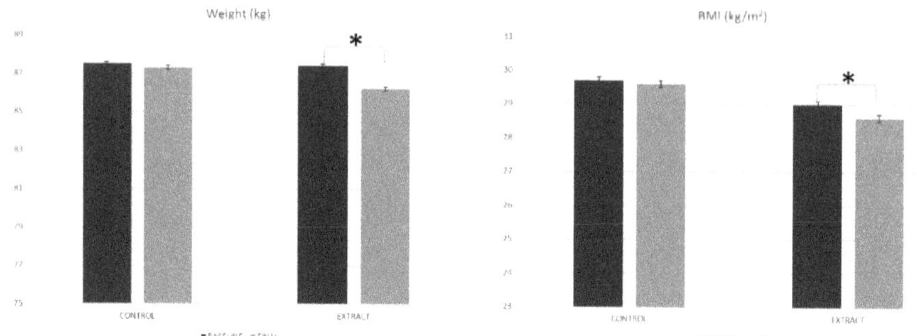

Figure 2. Weight and BMI evolution along the study. * Means statistically significant differences.

It should be as follows:

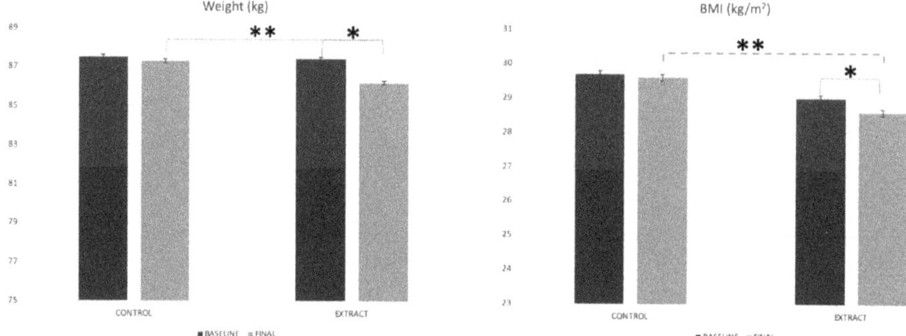

Figure 2. Weight and BMI evolution along the study. * Mean statistically significant differences. ** Mean statistically significant differences between groups.

Information is missing on Figure 4. The original figure is as follows:

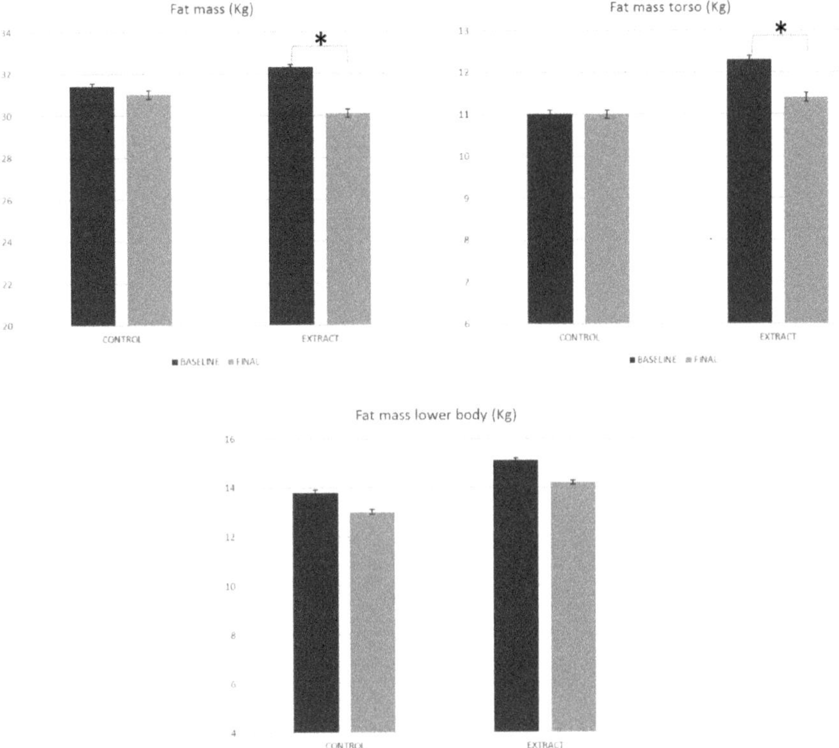

Figure 4. Body composition evolution along the study, measured by densitometry. * Means statistically significant differences.

It should be as follows instead:

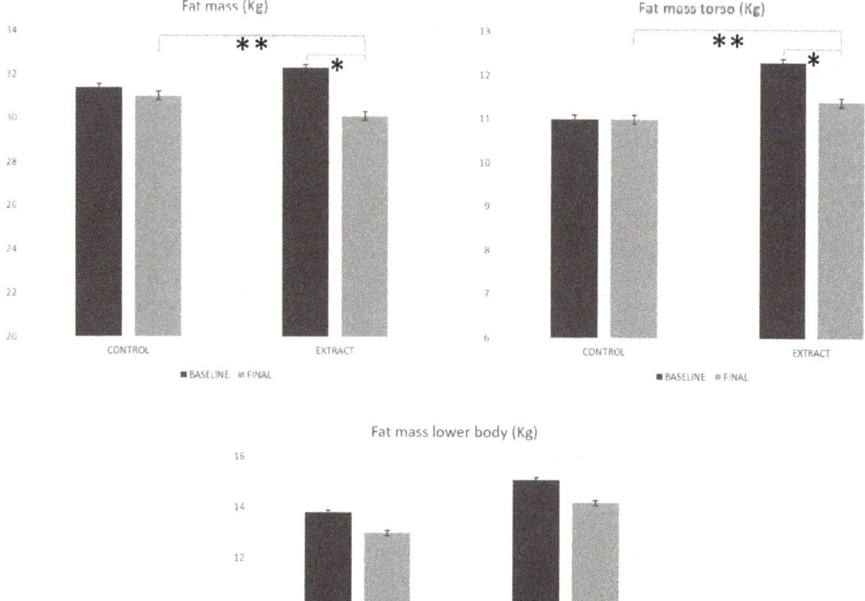

Figure 4. Body composition evolution along the study, measured by densitometry. * Means statistically significant differences. ** Mean statistically significant differences between groups.

These changes have no material impact on the conclusions of our paper.
The authors would like to apologize for any inconvenience caused to the readers by these changes.

Reference

1. Marhuenda, J.; Perez, S.; Victoria-Montesinos, D.; Abellán, M.S.; Caturla, N.; Jones, J.; López-Román, J. A Randomized, Double-Blind, Placebo Controlled Trial to Determine the Effectiveness a Polyphenolic Extract (Hibiscus sabdariffa and Lippia citriodora) in the Reduction of Body Fat Mass in Healthy Subjects. *Foods* **2020**, *9*, 55. [CrossRef] [PubMed]

Article

Polydatin and Resveratrol Inhibit the Inflammatory Process Induced by Urate and Pyrophosphate Crystals in THP-1 Cells

Francesca Oliviero [1,*,†], Yessica Zamudio-Cuevas [1,2,†], Elisa Belluzzi [1], Lisa Andretto [1], Anna Scanu [1], Marta Favero [1], Roberta Ramonda [1], Giampietro Ravagnan [3], Alberto López-Reyes [2], Paolo Spinella [4] and Leonardo Punzi [1,5]

[1] Rheumatology Unit, Department of Medicine—DIMED, University of Padova, 35128 Padova, Italy; yesszamudio@gmail.com (Y.Z.-C.); elisa.belluzzi@gmail.com (E.B.); lilli.1991@hotmail.it (L.A.); anna.scanu@unipd.it (A.S.); faveromarta@gmail.com (M.F.); roberta.ramonda@unipd.it (R.R.); punzileonardo@gmail.com (L.P.)

[2] Laboratorio de Líquido Sinovial, Instituto Nacional de Rehabilitación "Luis Guillermo Ibarra", 14380 Mexico City, Mexico; allorey@yahoo.com

[3] Institute of Translational Pharmacology—National Research Council, 00185 Rome, Italy; gprav@unive.it

[4] Clinical Nutrition Unit, Department of Medicine—DIMED, University of Padova, 35128 Padova, Italy; paolo.spinella@unipd.it

[5] Centre for Gout and Metabolic Bone and Joint Diseases, Rheumatology, SS Giovanni and Paolo Hospital, 30122 Venice, Italy

* Correspondence: francesca.oliviero@unipd.it; Tel.: +39-049-8218682

† These authors equally contributed to this article.

Received: 21 October 2019; Accepted: 5 November 2019; Published: 7 November 2019

Abstract: Resveratol (RES) and its natural precursor polydatin (PD) are polyphenols that may display a broad variety of beneficial effects including anti-inflammatory properties. This study aimed to investigate the role of RES and PD in the inflammatory process induced by monosodium urate (MSU) and calcium pyrophosphate (CPP) crystals in vitro. A monocytic cell line (THP-1) was primed for 3 hours with phorbol myristate acetate (100 ng/mL) and stimulated with synthetic MSU (0.05 mg/mL) and CPP (0.025 mg/mL) crystals. RES and PD were added to cultures concurrently with the crystals, or as 2-hour pretreatment. The effect of the two polyphenols was evaluated on intracellular and extracellular IL-1β levels, NACHT-LRRPYD-containing protein-3 (NLRP3) inflammasome expression, reactive oxygen species (ROS) and nitric oxide (NO) production, and the assessment of crystal phagocytosis. RES and PD strongly inhibited IL-1β induced by crystals after cell pretreatment. Cell pretreatment was effective also in reducing IL-1 mRNA expression while no effect was observed on NLRP3 gene expression. RES and PD had no effect on crystal phagocytosis when used as pretreatment. Both polyphenols were significantly effective in inhibiting ROS and NO production. Our results demonstrated that RES and PD are effective in inhibiting crystal-induced inflammation. Data obtained after cell pretreatment allow us to hypothesize that these polyphenols act on specific signaling pathways, preventing inflammation.

Keywords: polydatin; resveratrol; urate crystals; pyrophosphate crystals; crystal-induced inflammation

---

1. Introduction

The deposition of pathogenic crystals, such as calcium pyrophosphate (CPP) and monosodium urate (MSU) crystals in joints, triggers an important inflammatory reaction which leads to swelling, pain, limited joint function and, over time, tissue damage and disability. This process, which is characterized by an abnormal production of chemokines, cytokines, and reactive oxygen species (ROS) by the inflammatory cell infiltrate, can lead to chronic arthritis, namely, crystal-induced arthritis [1,2]. Since the time cytoplasmic NACHT-LRRPYD-containing protein-3 (NLRP3) inflammasome was first identified and its activation by MSU and CPP crystals was demonstrated [3], IL-1β has been considered the most important inflammatory mediator in crystal-induced inflammation, and it represents one of the main targets for new therapeutic approaches that have been or are being developed to treat crystal-related diseases [4–6].

Currently, corticosteroids, non-steroidal anti-inflammatory drugs (NSAIDs), and colchicine are effective drugs used to treat the acute attack in both gout, caused by MSU crystals, and pyrophosphate-related arthropathies, caused by CPP crystals [7]. However, while MSU crystal deposition can be prevented by treating hyperuricemia with urate-lowering therapy, there are no drugs able to prevent CPP crystal formation [6].

In recent years, significant interest has emerged in the beneficial health effects attributed to polyphenols. These plant-derived natural compounds have demonstrated a wide range of anti-inflammatory, anti-oxidant, and immunomodulant effects in many inflammatory chronic conditions [8–11].

Few years ago, we showed that the green tea polyphenol epigallocatechin-gallate affects the inflammatory response to CPP crystals by inhibiting chemokine release, chemotaxis, and interfering with membrane cell organization [12]. In the present study, we investigated the effect of a typical polyphenol of the Mediterranean diet, resveratrol (trans-3,5,4-trihydroxystilbene, RES) and its natural precursor, polydatin (3,5,4-trihydroxystilbene-3-O-beta-D-glucopyranoside, PD) in the in vitro model of crystal-induced inflammation. These compounds are mainly found in the roots of *Polygonum Cuspidatum*, grape, and red wine, and have shown a wide range of anti-inflammatory, anti-oxidant, and anti-tumoral properties [13–16].

However, only a few studies have evaluated the role of RES and PD in crystal-induced inflammation. While RES has shown to have an antihyperuricemic effect in mice [17,18] and to suppress inflammation through the inhibition of NLRP-3 inflammasome assembly [19], no studies have assessed the role of these polyphenols in the inflammatory process induced by CPP crystals.

We therefore assessed the influence of RES and PD on the IL-1β pathway and oxidative stress induced by crystals in THP-1 cells, suggesting an important role of these substances in preventing crystal-induced inflammation.

2. Materials and Methods

2.1. Reagents

THP-1 cells were purchased from the American Type Culture Collection®. CPP and MSU crystals were from InVivo Gen, Toulouse, France; phorbol myristate acetate (PMA), RPMI 1640, fetal calf serum, polymyxin B and phosphate buffer saline (PBS) were from Sigma-Aldrich, Milan, Italy.

2.2. Polyphenols

PD was extracted from *Polygonum Cuspidatum*, according to the procedure described in patent EP 1292320 B1 and kindly supplied by GLURES Srl (a spin-off of the National Research Council, Rome, Italy, purity >99%). RES was purchased from Sigma-Aldrich.

Both the compounds were dissolved in ethanol at 100 mM stock solution and diluted in culture medium prior to use.

2.3. Cell Culture and Treatment

THP-1 cells were cultured in RPMI 1640 medium supplemented with 10% heat inactivated fetal calf serum. Cells were primed for 3 h with PMA at 100 ng/mL and resuspended overnight with fresh medium supplemented with 10% fetal calf serum.

The following day, the cells were plated at a density of 50.000 cells/0.32 cm^2 and stimulated with sterile CPP or MSU crystals (cell culture tested and unable to activate TLR4-expressing cells) at the final concentration 0.025 and 0.05 mg/mL, respectively, and cultured for 24 h in presence of PD or RES [12]. Preliminarily dose–response experiments using different polyphenol concentrations ranging from 0.1 to 200 µM were conducted to determine the optimal RES (100 µM) and PD (200 µM) experimental concentrations.

Additional experiments have been performed after a 2 h pretreatment with PD and RES followed by their complete removal from the medium and subsequent stimulation with the crystals. Cells incubated with medium and polyphenols alone served as controls.

To exclude any contribution by endotoxin contamination, 10 mg/mL of polymyxin B was included in all the stimulation assays. All the experiments were performed three independent times.

2.4. Cell Viability, Death and Apoptosis Assays

Cell viability and proliferation were evaluated using the colorimetric MTT (3-(4,5-dimethylthiazol-2-yl)-2,5-diphenyltetrazolium bromide) mitochondrial activity assay (Chemicon) [20]. Cells were cultured in a 96-well plate and exposed to crystals, PD and RES for 24 h.

After removing the supernatant, cells were treated with the MTT solution according to the manufacturer's procedures and the absorbance intensity measured at 570 nm. The cell viability (%) was expressed as the percentage of cell survival in each group relative to the untreated control cells.

Cell death and apoptosis were assessed by image-based cytometry using the Alexa Fluor® 488 Annexin V/Dead Cell Apoptosis Kit (Molecular Probes).

2.5. IL-1β Measurement

Cell culture supernatants from different experimental conditions were examined for extracellular IL-1β concentration using Thermo Fischer Scientific ELISA kits (Waltham, MA, USA, detection limit 4 pg/mL). Intracellular IL-1β was determined in lysates obtained after three freeze–thaw cycles and resuspended in PBS [21].

2.6. Gene Expression Analysis

RNA extraction was carried out with the RNeasy mini kit (Qiagen, Milano, Italy) according to the manufacturer's instructions. RNA was treated with DNase (DNA-free kit, AMBION, Thermo Fisher Scientific, Waltham, MA, USA) to remove possible contaminating DNA and then reverse-transcribed using using a iScript™ cDNA Synthesis Kit (Bio-Rad, Milano, Italy) according to the manufacturer's instructions [22]. Quantitative real-time PCR (qPCR) was then carried out on a CFX96 Real-Time instrument (Bio-Rad), using Sso Fast Eva Green Supermix (Bio-Rad) according to the manufacturer's instructions. Levels of mRNA for each target gene were normalized to glyceraldehyde-3 phosphate dehydrogenase (GAPDH) and calculated according to the 2-ΔΔCt method [23]. The sequences of primers are reported in Table 1.

Table 1. Primer sets used for quantitative real-time PCR.

RNA Template	Forward Primer (5'-3')	Reverse Primer (5'-3')
GAPDH	CAAAGTTGTCATGGATGACC	CCATGGAGAAGGCTGGGG
IL-1β	CTGTCCTGCGTGTTGAAAGA	TTGGGTAATTTTTGGGATCTACA
ASC	CTGTCCATGGACGCCTTGG	CATCCGTCAGGACCTTCCCGT
NLRP3	CACCTGTTGTGCAATCTGAAG	GCAAGATCCTGACAACATGC

2.7. Assessment of Intracellular Oxidative Stress

Oxidative stress was evaluated through the determination of intracellular ROS by CellROX® Deep Red Reagent, a cell-permeable fluorogenic probe which exhibits excitation/emission at 640/665 nm upon oxidation [20]. The reagent was added at a concentration of 5 μM to the cells stimulated with crystals, RES and PD for the concentrations and indicated time, and incubated for 30 min at 37 °C.

The medium was then removed, and the cells washed three times with PBS. The fluorescence was measured using a fluorescence microscope (EVOS; FL Auto Cell Imaging System, Thermo Fisher, Waltham, MA, USA) and quantified using a Tali Image-based Cytometer (Thermo Fisher, Waltham, MA, USA) according to the manufacturer's instructions. Data analysis was performed based on fluorescence intensity expressed as arbitrary fluorescence units (AFU).

2.8. Nitric Oxide Determination

Intracellular NO was quantified using the commercial kit DAF-FM (4-amino-5-methylamino-2,7-difluorofluorescein diacetate, Molecular Probe), a pH-insensitive fluorescent dye that emits increased fluorescence after reaction with an active intermediate of NO formed during the spontaneous oxidation of NO to NO_2- at 515 nm. The fluorescence was quantified using a Tali Image-based Cytometer (Thermo Fisher, Waltham, MA, USA), according to the manufacturer's instructions. Data analysis was performed based on fluorescence intensities expressed as AFU.

2.9. Evaluation of CPP and MSU Crystal Phagocytosis

The phagocytosis of CPP and MSU crystals was assessed at different time points in the presence of RES and PD. Intracellular crystals were evaluated by means of ordinary and polarized light microscopy. The percentage of cells with internalized crystals was calculated on the total number of examined cells and expressed as a phagocytosis index.

2.10. Statistical Analysis

The data are expressed as the mean and standard deviation (SD) of three replicates. The significance between stimulation and treatment was calculated using the non-parametric Mann–Whitney test or the *t*-test when appropriate. Statistical analysis was performed with GraphPad Prism® 8 software. *p*-values less than 0.05 were considered significant.

3. Results

3.1. The Effect of PD and RES on Cell Viability, Death and Apoptosis Induced by Crystals

Preliminary experiments were conducted to assess cell toxicity of crystals and polyphenols. The MTT test showed a decrease in cell viability induced by CPP and MSU crystals which was abolished by PD and RES (supplementary material, Figure S1).

Cell stimulation with CPP and MSU crystals caused an increase in cell death (1.7- and 2.3-fold over the basal condition, respectively) and apoptosis (1.6- and 2-fold, respectively). Both these effects were suppressed by PD and RES (supplementary material, Figures S2 and S3).

3.2. The Effect of PD and RES on Intra- and Extra-cellular Levels of IL-1β

As shown in Figure 1 (panel A, B), RES inhibited extracellular levels of IL-1β induced by CPP ($p < 0.05$) and MSU crystals (non-significant) after 24 h, while PD showed an anti-inflammatory effect towards CPP only. Only RES was able to reduce IL-1β at the intra-cellular level. To evaluate whether the effect of PD and RES was mediated by a direct action on cell, THP-1 cells were pretreated with polyphenols and washed before crystal stimulation. Both PD and RES markedly reduced cytokine levels on culture medium in pretreated cells (Figure 1, panel C).

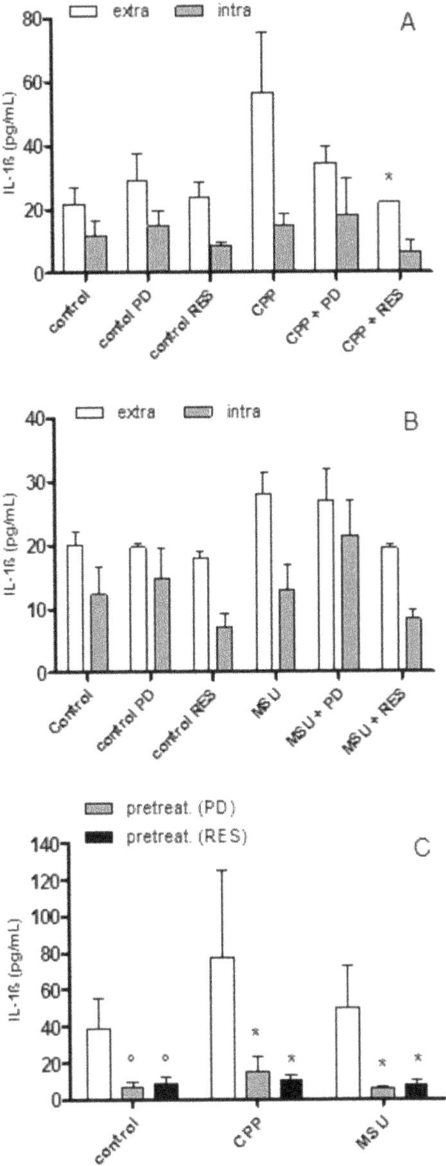

Figure 1. The effects of precursor polydatin (PD) and resveratol (RES) on calcium pyrophosphate (CPP) and monosodium urate (MSU) crystal-stimulated IL-1β production. Panel A, B: the cells were primed with PMA 100 ng/mL for 3 h and left overnight in fresh medium. The cells were then treated with CPP 0.025 mg/mL (**A**) or MSU 0.05 mg/mL (**B**) for 24 h in the presence of the two polyphenols. White bars show IL-1β extracellular concentrations, grey bars show IL-1β intracellular levels. In the left side of A and B figures are the control levels of IL-1β for each condition. Panel C: White bars show extracellular IL-1β level after crystal stimulation while grey and black bars show cytokine levels after 2 h cell pretreatment with PD (grey) and RES (black) followed by crystal stimulation. * $p < 0.05$ vs. CPP or MSU; ° $p < 0.05$ vs. basal control.

3.3. The Effect of PD and RES on IL-1β, NLRP3 and ASC Expression

Gene expression levels of IL-1β, an NLRP3 inflammasome and ASC, one of the inflammasome components, was examined after treatment and pretreatment with RES and PD.

As shown in Figure 2, CPP crystals induced a 1.7-fold change in IL-1β mRNA expression which was inhibited by RES only (panel A). By contrast, MSU crystals did not cause any change in IL-1β mRNA levels at the concentrations used in this study (panel B). The pretreatment of cells with PD and RES lead to a significant reduction of IL-1ß expression which was evident also at the basal conditions (panel A and B).

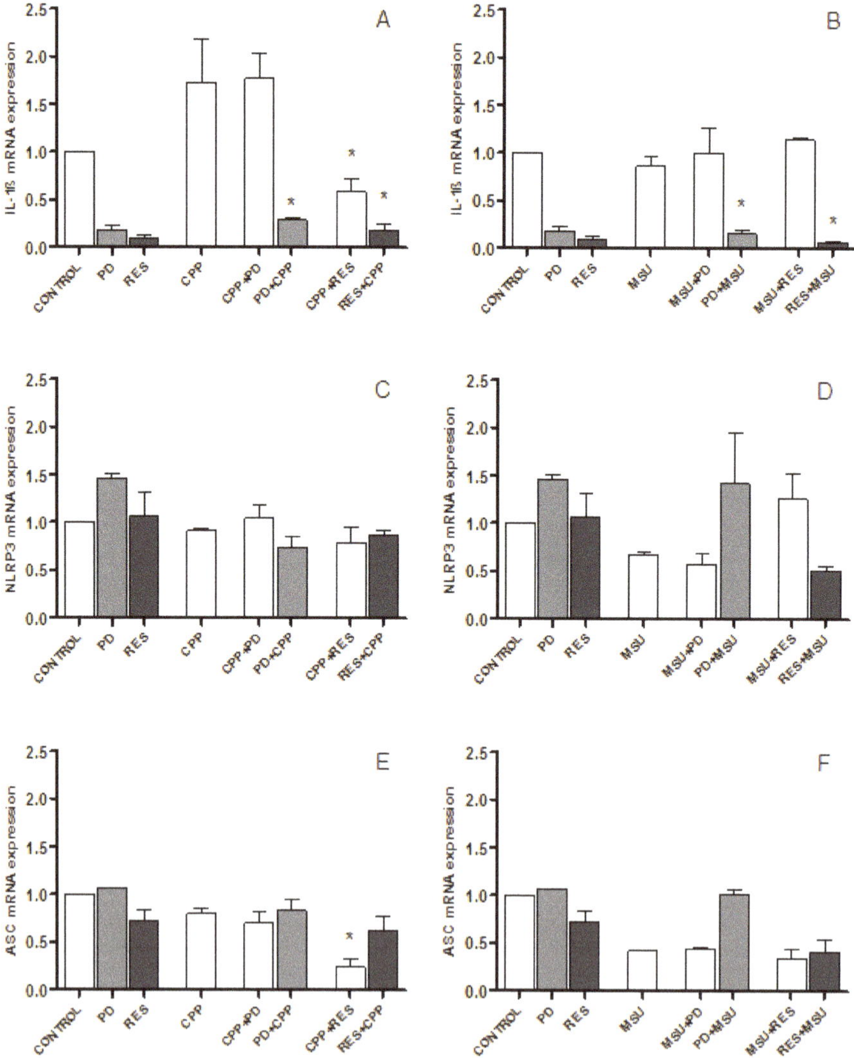

Figure 2. The effect of PD and RES on IL-1β, NLRP3 and ASC gene expression. Expression of IL-1β (**A,B**), NLRP3 (**C,D**) and ASC (**E,F**) mRNA in THP-1 cells stimulated with crystals and treated (white bars) or pretreated (gray bars) with PD and RES. * $p < 0.05$ vs. CPP or MSU.

There was, instead, no significant effect on NLRP3 and ASC gene expression by both crystals and polyphenols, although RES decreased ASC gene expression after stimulation with CPP crystals (Figure 2, panel C–F).

3.4. RES and PD Inhibit ROS Production Induced by Crystals

As ROS production has been demonstrated to play an important role in crystal-induced inflammation, the influence of the polyphenols on ROS released from treated and pretreated cells was investigated. THP-1 cells stimulated with CPP and MSU crystals for 24 h increased ROS production by approximately 2.4- and 5.5-fold compared with their basal level. Both RES and PD were significantly effective in inhibiting ROS production when added along with the crystals or 2 h before (Figure 3).

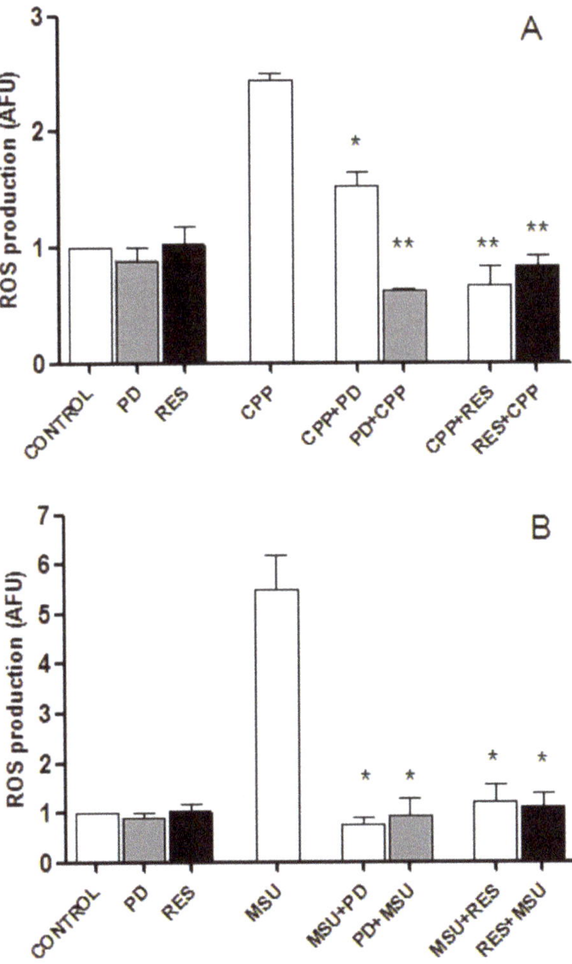

Figure 3. RES and PD suppress ROS production. White columns show ROS released by THP-1 cells treated with 0.025 mg/mL CPP (**A**) or 0.05 mg/mL MSU (**B**) crystals in presence of RES and PD. Grey and black columns indicate the amount of ROS released in cultures pretreated for 2 h with PD or RES, and treated with crystals after 24 h. On the left of the figures are the control conditions; CONTROL: cells only; PD: cells with PD only; RES, cells with RES only. Values are expressed as the mean ± standard deviation.* $p < 0.05$ vs. MSU or CPP.* $p < 0.05$, ** $p < 0.01$ vs. CPP or MSU.

3.5. RES and PD Inhibit Nitric oxide Production Induced by Crystals

CPP and MSU crystals induced a significant, although moderate, release of NO after 24 h stimulation with respect to control (Figure 4). In the presence of RES and PD, the amount of NO returned to basal levels. Similarly, the pretreatment of cells with the two polyphenols inhibited the production of NO in culture supernatants.

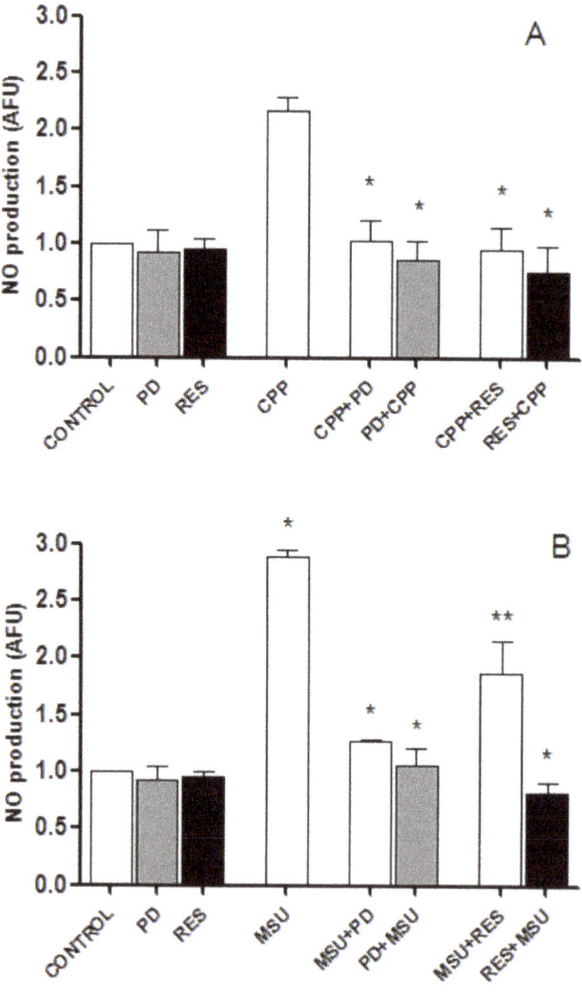

Figure 4. RES and PD decrease NO production. White columns show quantification of NO production by THP-1 cells treated with 0.025 mg/mL CPP (**A**) or 0.05 mg/mL MSU (**B**) crystals in presence of RES and PD. Grey and black columns indicate the amount of NO released in cultures pretreated for 2 h with PD or RES, and treated with crystals after 24 h. On the left of the figures are the control conditions; CONTROL: cells only; PD: cells with PD only; RES, cells with RES only. Values are expressed as the mean ± standard deviation.* $p < 0.01$ vs. MSU or CPP; ** $p < 0.05$ vs. MSU.

3.6. The Effect of RES and PD on Crystal Phagocytosis

As crystal phagocytosis represents a fundamental step in triggering the inflammatory response to crystals, the hypothesis that PD and RES could interfere with crystal internalization has been examined.

The phagocytosis was measured at different time points (4, 24, and 48 h) on cells treated or pretreated with PD and RES. Figure 5 shows the variation of the phagocytosis index over time. While PD reduced in a non-significant manner the uptake of crystals, the phagocytosis index diminished significantly in presence of RES. However, the pretreatment of the cells with the two polyphenols had no effect on the phagocytosis process.

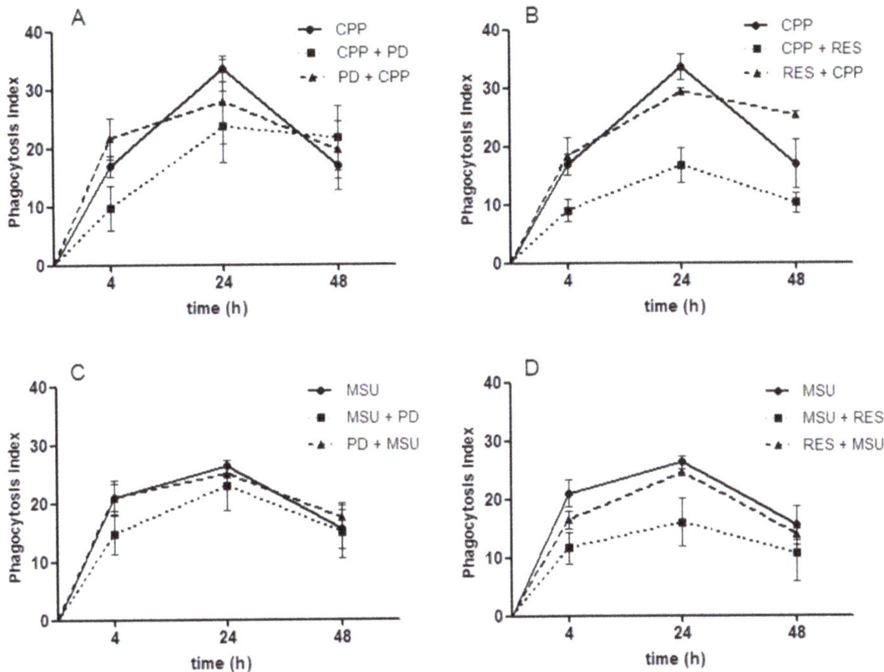

Figure 5. The effect of PD and RES on CPP and MSU crystal phagocytosis. (**A**) cells stimulated with CPP crystals and treated (square) or pretreated (triangle) with PD; (**B**) cells stimulated with CPP crystals and treated (square) or pretreated (triangle) with RES; (**C**) cells stimulated with MSU crystals and treated (square) or pretreated (triangle) with PD; (**D**) cells stimulated with MSU crystals and treated (square) or pretreated (triangle) with RES. Phagocytosis was evaluated using ordinary/polarized light microscopy assessing the presence of intracellular crystals at the indicated time points. Crystals were used at the concentration of 0.025 (CPP) and 0.05 mg/mL (MSU). * $p < 0.05$ vs. CPP.

4. Discussion

This study showed that RES and PD, two important polyphenols found in grape juice and red wine [24], reduce IL-1β release in THP-1 cells stimulated by pathogenic crystals affecting inflammasome activity but not its gene expression. We have found that these compounds influence the production of ROS and NO and show a strong preventing effect towards the release of these inflammatory mediators. We also demonstrated that RES had a stronger anti-inflammatory effect with respect to PD, explained, at least in part, by its capacity to inhibit the process of crystal phagocytosis.

In humans, CPP and MSU crystal deposition triggers powerful inflammatory reactions causing, respectively, calcium pyrophosphate crystal-associated arthritis (pseudogout) and gout. Both types of crystals activate specific signaling pathways leading to the production of inflammatory cytokines, chemokines, and pro-oxidant molecules in the synovial compartment [25]. As stated above, IL-1β is one of the most important mediators in crystal-induced inflammation [3,5].

It is known that a co-stimulus is needed to prime cells to generate IL-1β when exposed to pathogenic crystals. It has been hypothesized that serum proteins [26,27] and Toll-like receptor ligands such as free fatty acids [28] synergize with crystals on IL-1β induction. In this study, we used the protein kinase C activator PMA which influences cell differentiation and stimulates monocyte functions. Using this model, we previously demonstrated that the most important catechin of green tea, epigallocatechin-gallate, leads to a significant inhibition of the inflammatory response triggered by CPP crystals [12]. In the present study, we observed an increase of IL-1β production induced by crystals after cell priming, which was inhibited by RES and PD both at intracellular and extracellular levels. The inflammatory effect of CPP in this cellular model was more pronounced with respect to that induced by MSU crystals.

CPP, but not MSU, crystals induced an increase in IL-1β mRNA levels which was inhibited only by RES. By contrast, the pretreatment of cells with PD and RES caused a significant reduction of IL-1β gene expression which was evident also at the basal conditions. There was, instead, no significant effect on NLRP3 and ASC gene expression by both crystals and polyphenols, although RES diminished ASC gene expression after treatment with CPP crystals. Therefore, the results showed that the decrease in IL-1β observed in THP-1 cells stimulated with crystals and treated with RES and PD was associated with a decrease in IL-1β gene expression but not in NLRP3 and ASC gene expression.

Recently, Yang et al. showed that RES decreases IL-1β levels in peripheral blood mononuclear cells obtained from patients with gout and challenged with MSU crystals. According to their results, RES was able to restore the levels of sirtuin-1 protein previously decreased by crystals [29]. In that study, RES showed opposite effects on pro-IL-1β and the mature form, depending on the concentration of the polyphenol.

RES, along with red wine extracts, has been shown to decrease IL-1β secretion in murine macrophage prior to cell priming with LPS [30]. In particular, the authors showed that RES was able to decrease IL-1β production and NLRP3 expression after pretreatment, and demonstrated that the effect of this polyphenol on the IL-1 activation pathway can differ considerably depending on the NLRP3 activator used.

In line with that study, our results showed that the pretreatment of cells with the two polyphenols almost completely abrogated the inflammatory response, suggesting a major role on cell priming rather than on the activation signal. Of note, PD and RES were removed from the cultures after a 2 h pretreatment, suggesting a sustained anti-inflammatory effect over time. Cell pretreatment conditioned the basal state of the cells, making them less prone to stimulation.

We then investigated whether RES and PD were able to affect oxidative and nitrosative stress which has been demonstrated to be induced by crystals in synoviocytes, chondrocytes, and macrophages [20,31,32]. In our experiments, RES and PD decreased ROS and NO production, and the effect was more effective when cells were pretreated with the two compounds.

These effects could be related to the affinity of these compounds for the lipid part of membranes interacting with the head groups of phospholipids [33]. This mechanism has been postulated to protect lipid membranes from oxidative stress and hydrolytic attack [33].

The importance of this lipid protective effect has recently emerged by using PD as topical applications to prevent drug-induced skin toxicities [34].

We finally investigated whether PD and RES could interfere with crystal phagocytosis, which represents a fundamental step in triggering the inflammatory response to crystals. We observed that RES significantly inhibited CPP and MSU crystal internalization when added concurrently with the crystals but they had no effect after cell pretreatment. The influence of RES on phagocytosis has been described in different cellular models. It has been demonstrated to inhibit bacteria phagocytosis in THP-1 cells [35] and the phagocytic activity of microglia [36] as well as to enhance the phagocytosis of yeast in human macrophage-like cells [37]. In THP-1 cells, this effect has been observed to be mediated by gene expression inhibition of the phagocytic scavenger receptors and C-type lectin receptors carried out by RES [35].

In our conditions, it could be hypothesized that RES acts on innate immunity surface receptors, such as TLRs, whose ligands serve to prime cells in the signaling transduction pathway activated by crystals. However, TLR-2 and TLR-4 inhibition by monoclonal antibodies does not affect the rate of CPP and MSU crystals phagocytosis in this model (personal experimental observations). Whether RES acts on other types of receptors will be investigated in further studies.

As far as PD is concerned, it showed only a modest effect on phagocytosis. With respect to RES, PD has a glucoside group which seems to confer to the molecule's different biological properties [38]. Among these, it has been demonstrated that PD, unlike RES which penetrates the cell passively, enters the cell via an active mechanism using glucose carriers [39].

In our model, PD showed a smaller effect on IL-1 inhibition compared to RES but a strong anti-oxidant effect towards the release of ROS and NO.

A limitation of this study might concern the concentration of RES used in this study, which, although in line with other in vitro studies, is probably too high to be reached in the bloodstream after dietary RES consumption. However, it is possible to achieve such plasmatic concentration by administering resveratrol supplements [40]. Furthermore, the conjugated glucuronides and sulfate metabolites of RES retain biological activity [41].

5. Conclusions

In conclusion, our study demonstrated that PD, and to a major extent RES, have a strong preventive anti-inflammatory effect in THP-1 cells primed with PMA and stimulated with pathogenic crystals. Whether their mechanism of action is related to the interaction of these compounds with cell membranes and to the protection of cells from oxidative reactions, might represent a future focus of research.

Finally, given the auto-inflammatory nature of crystal-induced arthritis with acute attacks which resolve spontaneously, dietary supplementation with these polyphenols could support pharmacological treatment in patients and prevent the acute phase of the disease. Clinical studies assessing the benefits of RES and PD in the management of gout and pseudogout as well as in reducing biological inflammatory biomarkers could provide further data on the anti-inflammatory preventive effect of these polyphenols and their possible application in those diseases.

Supplementary Materials: The following are available online at http://www.mdpi.com/2304-8158/8/11/560/s1, Figure S1. The effect of PD and RES on cell viability induced by crystals, Figure S2. The effect of PD and RES on cell death induced by crystals, Figure S3. The effect of PD and RES on cell apoptosis induced by crystals.

Author Contributions: Conceptualization, F.O., Y.Z.-C., G.R., P.S., and L.P.; methodology, F.O., Y.Z.-C., E.B., L.A., and A.S.; formal analysis, F.O., and Y.Z.-C.; investigation, F.O., Y.Z.-C., E.B., L.A., and A.S.; resources, F.O., L.P. and A.L.-R.; data curation, F.O., Y.Z.-C.; writing—original draft preparation, F.O. and Y.Z.-C.; writing—review and editing, E.B., M.F., R.R., G.R., A.L.-R., P.S., and L.P.; supervision, R.R., G.R., P.S., L.P. and A.L.-R.; project administration, F.O., Y.Z.-C.; funding acquisition, F.O., A.L.-R., L.P.

Funding: This research was funded by the UNIVERSITY OF PADOVA, grant number DOR1872371/18, DOR1972490/19 and the ITALIAN MINISTRY OF EDUCATION, UNIVERSITY AND RESEARCH, grant number FFABR2017.

Conflicts of Interest: The authors declare no conflict of interest.

References

1. Mahon, O.R.; Dunne, A. Disease-Associated Particulates and Joint Inflammation; Mechanistic Insights and Potential Therapeutic Targets. *Front. Immunol.* **2018**, *9*, 1145. [CrossRef]
2. Zamudio-Cuevas, Y.; Hernández-Díaz, C.; Pineda, C.; Reginato, A.M.; Cerna-Cortés, J.F.; Ventura-Ríos, L.; López-Reyes, A. Molecular basis of oxidative stress in gouty arthropathy. *Clin. Rheumatol.* **2015**, *34*, 1667–1672. [CrossRef]
3. Martinon, F.; Pétrilli, V.; Mayor, A.; Tardivel, A.; Tschopp, J. Gout-associated uric acid crystals activate the NALP3 inflammasome. *Nature* **2006**, *440*, 237–241. [CrossRef]
4. Dinarello, C.A. An expanding role for interleukin-1 blockade from gout to cancer. *Mol. Med.* **2014**, *20*, S43–S58. [CrossRef] [PubMed]

5. So, A.; Dumusc, A.; Nasi, S. The role of IL-1 in gout: From bench to bedside. *Rheumatology* **2018**, *57* (Suppl. 1), i12–i19. [PubMed]

6. Andrés, M.; Sivera, F.; Pascual, E. Therapy for CPPD: Options and Evidence. *Curr. Rheumatol. Rep.* **2018**, *19*, 31. [CrossRef] [PubMed]

7. Stamp, L.K.; Merriman, T.R.; Singh, J.A. Expert opinion on emerging urate-lowering therapies. *Expert. Opin. Emerg. Drugs* **2018**, *23*, 201–209. [CrossRef] [PubMed]

8. Ding, S.; Jiang, H.; Fang, J. Regulation of Immune Function by Polyphenols. *J. Immunol. Res.* **2018**, *2018*, 1264074. [CrossRef]

9. Oliviero, F.; Scanu, A.; Zamudio-Cuevas, Y.; Punzi, L.; Spinella, P. Anti-inflammatory effects of polyphenols in arthritis. *J. Sci. Food. Agric.* **2018**, *98*, 1653–1659. [CrossRef]

10. Billingsley, H.E.; Carbone, S. The antioxidant potential of the Mediterranean diet in patients at high cardiovascular risk: An in-depth review of the PREDIMED. *Nutr. Diabetes* **2018**, *8*, 13. [CrossRef]

11. Cory, H.; Passarelli, S.; Szeto, J.; Tamez, M.; Mattei, J. The Role of Polyphenols in Human Health and Food Systems: A Mini-Review. *Front. Nutr.* **2018**, *5*, 87. [CrossRef] [PubMed]

12. Oliviero, F.; Sfriso, P.; Scanu, A.; Fiocco, U.; Spinella, P.; Punzi, L. Epigallocatechin-3-gallate reduces inflammation induced by calcium pyrophosphate crystals in vitro. *Front. Pharmacol.* **2013**, *4*, 51. [CrossRef] [PubMed]

13. Ravagnan, G.; De Filippis, A.; Cartenì, M.; De Maria, S.; Cozza, V.; Petrazzuolo, M.; Tufano, M.A.; Donnarumma, G. Polydatin, a natural precursor of resveratrol, induces β-defensin production and reduces inflammatory response. *Inflammation* **2013**, *36*, 26–34. [CrossRef] [PubMed]

14. de Sá Coutinho, D.; Pacheco, M.T.; Frozza, R.L.; Bernardi, A. Anti-Inflammatory Effects of Resveratrol: Mechanistic Insights. *Int. J. Mol. Sci.* **2018**, *19*, 1812. [CrossRef] [PubMed]

15. Di Benedetto, A.; Posa, F.; De Maria, S.; Ravagnan, G.; Ballini, A.; Porro, C.; Trotta, T.; Grano, M.; Muzio, L.L.; Mori, G. Polydatin, Natural Precursor of Resveratrol, Promotes Osteogenic Differentiation of Mesenchymal Stem Cells. *Int. J. Med. Sci.* **2018**, *15*, 944–952. [CrossRef] [PubMed]

16. Martano, M.; Stiuso, P.; Facchiano, A.; De Maria, S.; Vanacore, D.; Restucci, B.; Rubini, C.; Caraglia, M.; Ravagnan, G.; Lo Muzio, L. Aryl hydrocarbon receptor, a tumor grade-associated marker of oral cancer, is directly downregulated by polydatin: A pilot study. *Oncol. Rep.* **2018**, *40*, 1435–1442. [CrossRef] [PubMed]

17. Shi, Y.W.; Wang, C.P.; Liu, L.; Liu, Y.L.; Wang, X.; Hong, Y.; Li, Z.; Kong, L.D. Antihyperuricemic and nephroprotective effects of resveratrol and its analogues in hyperuricemic mice. *Mol. Nutr. Food. Res.* **2012**, *56*, 1433–1444. [CrossRef]

18. Chen, H.; Zheng, S.; Wang, Y.; Zhu, H.; Liu, Q.; Xue, Y.; Qiu, J.; Zou, H.; Zhu, X. The effect of resveratrol on the recurrent attacks of gouty arthritis. *Clin. Rheumatol.* **2016**, *35*, 1189–1195. [CrossRef]

19. Misawa, T.; Saitoh, T.; Kozaki, T.; Park, S.; Takahama, M.; Akira, S. Resveratrol inhibits the acetylated α-tubulin-mediated assembly of the NLRP3-inflammasome. *Int. Immunol.* **2015**, *27*, 425–434. [CrossRef]

20. Zamudio-Cuevas, Y.; Martínez-Flores, K.; Fernández-Torres, J.; Loissell-Baltazar, Y.A.; Medina-Luna, D.; López-Macay, A.; Camacho-Galindo, J.; Hernández-Díaz, C.; Santamaría-Olmedo, M.G.; López-Villegas, E.O.; et al. Monosodium urate crystals induce oxidative stress in human synoviocytes. *Arthritis Res. Ther.* **2016**, *18*, 117. [CrossRef]

21. Cleophas, M.C.; Crişan, T.O.; Lemmers, H.; Toenhake-Dijkstra, H.; Fossati, G.; Jansen, T.L.; Dinarello, C.A.; Netea, M.G.; Joosten, L.A. Suppression of monosodium urate crystal-induced cytokine production by butyrate is mediated by the inhibition of class I histone deacetylases. *Ann. Rheum. Dis.* **2016**, *75*, 593–600. [CrossRef] [PubMed]

22. Belluzzi, E.; Olivotto, E.; Toso, G.; Cigolotti, A.; Pozzuoli, A.; Biz, C.; Trisolino, G.; Ruggieri, P.; Grigolo, B.; Ramonda, R.; et al. Conditioned media from human osteoarthritic synovium induces inflammation in a synoviocyte cell line. *Connect. Tissue Res.* **2019**, *60*, 136–145. [CrossRef] [PubMed]

23. Livak, K.J.; Schmittgen, T.D. Analysis of relative gene expression data using real-time quantitative PCR and the 2(-Delta Delta C (T)) Method. *Methods* **2001**, *25*, 402–408. [CrossRef] [PubMed]

24. Xia, E.Q.; Deng, G.F.; Guo, Y.J.; Li, H.B. Biological activities of polyphenols from grapes. *Int. J. Mol. Sci.* **2010**, *11*, 622–646. [CrossRef] [PubMed]

25. Busso, N.; Ea, H.K. The mechanisms of inflammation in gout and pseudogout (CPP-induced arthritis). *Reumatismo* **2012**, *63*, 230–237. [CrossRef] [PubMed]

26. Oliviero, F.; Scanu, A.; Dayer, J.M.; Fiocco, U.; Sfriso, P.; Punzi, L. Response to 'Plasma proteins present in osteoarthritic synovial fluid can stimulate cytokine production via Toll-like receptor 4'. *Arthritis Res. Ther.* **2012**, *14*, 405. [CrossRef]

27. Scanu, A.; Oliviero, F.; Gruaz, L.; Galozzi, P.; Luisetto, R.; Ramonda, R.; Burger, D.; Punzi, L. Synovial fluid proteins are required for the induction of interleukin-1β production by monosodium urate crystals. *Scand. J. Rheumatol.* **2016**, *45*, 384–393. [CrossRef]

28. Mylona, E.E.; Mouktaroudi, M.; Crisan, T.O.; Makri, S.; Pistiki, A.; Georgitsi, M.; Savva, A.; Netea, M.G.; van der Meer, J.W.; Giamarellos-Bourboulis, E.J.; et al. Enhanced interleukin-1β production of PBMCs from patients with gout after stimulation with Toll-like receptor-2 ligands and urate crystals. *Arthritis Res. Ther.* **2012**, *14*, R158. [CrossRef]

29. Yang, Q.B.; He, Y.L.; Zhong, X.W.; Xie, W.G.; Zhou, J.G. Resveratrol ameliorates gouty inflammation via upregulation of sirtuin 1 to promote autophagy in gout patients. *Inflammopharmacology* **2019**, *27*, 47–56. [CrossRef]

30. Chalons, P.; Amor, S.; Courtaut, F.; Cantos-Villar, E.; Richard, T.; Auger, C.; Chabert, P.; Schni-Kerth, V.; Aires, V.; Delmas, D. Study of Potential Anti-Inflammatory Effects of Red Wine Extract and Resveratrol through a Modulation of Interleukin-1-Beta in Macrophages. *Nutrients* **2018**, *10*, 1856. [CrossRef]

31. Peng, Y.J.; Lee, C.H.; Wang, C.C.; Salter, D.M.; Lee, H.S. Pycnogenol attenuates the inflammatory and nitrosative stress on joint inflammation induced by urate crystals. *Free Radic. Biol. Med.* **2012**, *52*, 765–774. [CrossRef] [PubMed]

32. Kim, S.K.; Choe, J.Y.; Park, K.Y. Rebamipide Suppresses Monosodium Urate Crystal-Induced Interleukin-1β Production Through Regulation of Oxidative Stress and Caspase-1 in THP-1 Cells. *Inflammation* **2016**, *39*, 473–482. [CrossRef] [PubMed]

33. Fei, Q.; Kent, D.; Botello-Smith, W.M.; Nur, F.; Nur, S.; Alsamarah, A.; Chatterjee, P.; Lambros, M.; Luo, Y. Mechanism of Resveratrol's Lipid Membrane Protection. *Sci. Rep.* **2018**, *8*, 1587. [CrossRef] [PubMed]

34. Fuggetta, M.P.; Migliorino, M.R.; Ricciardi, S.; Osman, G.; Iacono, D.; Leone, A.; Lombardi, A.; Ravagnan, G.; Greco, S.; Remotti, D.; et al. Prophylactic Dermatologic Treatment of Afatinib-Induced Skin Toxicities in Patients with Metastatic Lung Cancer: A Pilot Study. *Scientifica* **2019**, *2019*, 9136249. [CrossRef] [PubMed]

35. Iyori, M.; Kataoka, H.; Shamsul, H.M.; Kiura, K.; Yasuda, M.; Nakata, T.; Hasebe, A.; Shibata, K. Resveratrol modulates phagocytosis of bacteria through an NF-kappaB-dependent gene program. *Antimicrob. Agents Chemother.* **2008**, *52*, 121–127. [CrossRef]

36. Park, E.; Kim, D.K.; Chun, H.S. Resveratrol Inhibits Lipopolysaccharide-induced Phagocytotic Activity in BV2 cells. *J. Korean Soc. Appl. Biol. Chem.* **2012**, *55*, 803–807. [CrossRef]

37. Bertelli, A.A.; Ferrara, F.; Diana, G.; Fulgenzi, A.; Corsi, M.; Ponti, W.; Ferrero, M.E.; Bertelli, A. Resveratrol, a natural stilbene in grapes and wine, enhances intraphagocytosis in human promonocytes: A co-factor in antiinflammatory and anticancer chemopreventive activity. *Int. J. Tissue React.* **1999**, *21*, 93–104.

38. Fabris, S.; Momo, F.; Ravagnan, G.; Stevanato, R. Antioxidant properties of resveratrol and piceid on lipid peroxidation in micelles and monolamellar liposomes. *Biophys. Chem.* **2008**, *135*, 76–83. [CrossRef]

39. Romero-Perez, A.I.; Ibern-Gomez, M.; Lamuela-Raventos, R.M.; de La Torre-Boronat, M.C. Piceid, the major resveratrol derivative in grape juices. *J. Agric. Food. Chem.* **1999**, *47*, 1533–1536. [CrossRef]

40. Walle, T.; Hsieh, F.; DeLegge, M.H.; Oatis, J.E., Jr.; Walle, U.K. High absorption but very low bioavailability of oral resveratrol in humans. *Drug Metab. Dispos.* **2004**, *32*, 1377–1382. [CrossRef]

41. Springer, M.; Moco, S. Resveratrol and Its Human Metabolites-Effects on Metabolic Health and Obesity. *Nutrients* **2019**, *11*, 143. [CrossRef] [PubMed]

Review

Cognitive Function and Consumption of Fruit and Vegetable Polyphenols in a Young Population: Is There a Relationship?

Juan Ángel Carrillo, M Pilar Zafrilla and Javier Marhuenda *

Faculty of Health Sciences, Universidad Católica de San Antonio, 30107 Murcia, Spain; jacarrillo4@alu.ucam.edu (J.Á.C.); mpzafrilla@ucam.edu (M.P.Z.)
* Correspondence: jmarhuenda@ucam.edu; Tel.: +34-968-27-86-18

Received: 27 August 2019; Accepted: 2 October 2019; Published: 17 October 2019

Abstract: Scientific evidence has shown the relationship between consumption of fruits and vegetables and their polyphenols with the prevention or treatment of diseases. The aim of this review was to find out whether the same relationship exists between fruits and vegetables and cognitive function, especially memory, in a young population. The mechanisms by which polyphenols of fruits and vegetables can exert cognitive benefits were also evaluated. These compounds act to improve neuronal plasticity through the protein CREB (Camp Response Element Binding) in the hippocampus, modulating pathways of signaling and transcription factors (ERK/Akt). In the same way, brain-derived neurotrophic factor (BDNF) is implicated in the maintenance, survival, growth, and differentiation of neurons. All these effects are produced by an increase of cerebral blood flow and an increase of the blood's nitric oxide levels and oxygenation.

Keywords: cognitive function; polyphenols; flavonoids; CREB protein; BDNF; memory; fruits and vegetables; cerebral blood flow

---

1. Impact of the Consumption of Fruit and Vegetables on Health

A healthy lifestyle and nutritional habits are essential for the maintenance of a healthy state in a young population. These habits should include adequate intake of fruits and vegetables [1,2], habitual consumption of omega series fatty acids (about 250–500 mg/day of 6-3/Ω) [3], and limiting the consumption of saturated fatty acids over 10%. Apart from physical activity, adequate intake of vitamins, minerals, and other bioactive compounds through the consumption of fruits and vegetables is essential. In fact, consumption of these foods is related to better aging conditions and a lowered risk of chronic diseases, cancer, physical dysfunction, and mental illness [4,5]. One of the dietary patterns most in line with these guidelines is the Mediterranean diet, characterized by its high content of antioxidants and polyphenols [6–9]. Recent publications have revealed the association of grain, fruit, and vegetable intake with a decreased risk of cognitive impairment [10].

Consumption of fruits and vegetables in adequate amounts has a direct protective impact on health and is often included in the treatment plans of certain diseases such as obesity [11]; chronic degenerative diseases [12,13]; cardiovascular diseases, through the regulation of cholesterol [14]; the decrease of lipid cell damage in leukocytes and erythrocytes [15]; and functional recovery after stroke and enhanced cognitive function in a parallel block-randomized clinical trial [16]. These impacts have been demonstrated in a recent umbrella review of observational studies [17] and in studies with mice for cerebral ischemia-reperfusion injuries [18].

Several recent meta-analyses have provided evidence supporting a positive relationship between the consumption of fruits and vegetables and prevention of type 2 diabetes mellitus [19], cancer risk [20], and mortality in general [21].

However, other authors have found evidence of this relationship between fruit and vegetable intake and health with only some diseases, such as colon and rectal cancer, hip fracture, stroke, depression, and pancreatic diseases, and limited associations with other outcomes [17]. Bioactive compounds present in fruits and vegetables have a great capacity to neutralize free radicals, averting the deterioration of aging and degenerative diseases. The main proven effects are decreasing oxidative injury [22], and exerting anti-inflammatory and neuroprotective effects [23,24].

Bioactive compounds vary for different fruits and vegetables. For example, pomegranate shows a higher antioxidant capacity than apple [25], presenting similar values to different types of berries, especially members of the Rosaceae and Ericaceae families. The bioactive compounds in berries contain mainly phenolic acids, flavonoids, such as anthocyanins and flavonols, and tannins [26,27]. The effect of bioactive compounds also depends on the size of the portion consumed [28] and the frequency of consumption, as reported in a 12-week study in humans. In this case, the result was significant for memory and cognitive benefits in adults [29].

The mechanisms of action of bioactive compounds have been reported in different studies. Some authors consider that metabolites pass to the colon, where they are metabolized by the action of the local microbiota, leading to small phenolic acids and aromatic catabolites that are absorbed into the circulatory system [30–33]. In all cases, the determined health benefits must be proved by well-controlled, in vivo intervention studies [34].

The young population has significant neuronal plasticity, showing a high rate of cellular replacement. For this reason, their dietary patterns include foods which have high contents of flavonoids, in order to improve such capacity. Moreover, age cognitive impairment is averted in these cases [35–37].

In the elderly population, the susceptibility to oxidative damage is especially increased in the brain. Similarly to young adults, the consumption of flavonoids averts the damage caused by aging. One of the mechanisms responsible might be the flavonoids' ability to increase cellular signaling and neuronal communication. A high daily intake of fruits and vegetables leads to higher antioxidant levels, lower levels of biomarkers of oxidative stress, and better cognitive performance than healthy subjects consuming low amounts of fruits and vegetables [13,38].

Thus, flavonoids are considered important in cognitive function, and although the mechanisms of action have not yet been clarified, it is known that these compounds modulate cerebral blood flow, inducing changes in processing of memory.

2. Polyphenols

Polyphenols are widely distributed in fruits and vegetables and other dietary sources such as nuts, whole grains, olive oil, and infused beverages such as tea. The most common classification of polyphenols is according to the presence or absence of a flavonoid skeleton, classifying them into flavonoids and non-flavonoids [31,39,40].

Flavonoids include anthocyanins [41], which are major polyphenols in berries and red fruits such as grapes [23], catechins [42], proanthocyanidins, flavonols, highly present in green tea and cocoa [43,44], and stilbenes such as resveratrol [45,46]. Other flavonoids are quercetin and kaempferol, which are present in onions, leeks or broccoli. Flavones can be found in high concentrations in parsley or celery, while isoflavones are major polyphenols of soy. Moreover, hesperidin is a major compound in citrus fruits, and cereals such as sorghum [47].

Protective Capacity of Polyphenols

Polyphenols are able to reduce oxidative stress and cell damage [48,49]. Consistent with this finding, a study where rats were fed with blueberries [50] showed a decrease of oxidative DNA damage, along with lipid peroxidation which was measured by F2-isoprostanes and an increase of reduced glutathione. On the other hand, in 2012, another study reported a similar effect with vegetable soup with a high content of carotenoids. In that case, the results were not the same regarding oxidative damage, except the levels of glutathione, which were also reduced after four weeks of consumption [51].

In chronic inflammatory processes, there is a loss of barrier effect and an overproduction of oxidants such as eicosanoids, cytokines, chemokines, and metalloproteins. The neuroinflammatory process is carried by activated resident glial cells (astrocytes and microglia) and determined by accumulation of circulating immune cells and releasing of proinflammatory cytokines, including tumor necrosis factor (TNF)-α, IL-6, nitric oxide, and reactive oxygen species (ROS) [52]. Polyphenols work to inhibit the release of cytokines, inhibit NADPH oxidase activation and consequent ROS generation in activated glia, and modulate the response to predisposition to chronic inflammation, such as in rheumatoid arthritis and psoriatic diseases. The levels of these mediators contribute to clinical symptoms [53] and modulate cellular signaling processes such as cellular growth and differentiation mediated IL-8 and TNFα through the activation of NF-kB and AP-1 transcription factors.

In neurodegenerative diseases, the presence of those inflammatory biomarkers is related to cognitive impairment [54–56] and osteoarthritis by mediation of nitric oxide synthase, cyclo-oxygenase, and metalloproteinases [57]. That prevention of neuroinflammation related to polyphenols has been confirmed with isoflavones recently, and these can cross the blood–brain barrier, making them desirable agents for the prevention of neurological symptoms by inhibiting proinflammatory cytokines, such as TNF-α and in IL-6, mitigating microglial activation, and preventing neuroinflammation. [58]. This inflammatory response has also been studied in muscle damage and homeostasis in intense physical resistance exercise. In a recent study, it was demonstrated to have an effect on post-race muscle damage and inflammation attenuation before and after a marathon race [59]. Moreover, this effect was also reported in young adults after the consumption of vitamin C present in fruits and vegetables through measurement of plasmatic inflammatory markers [60].

Finally, a cross-sectional study revealed the benefits related to the intake of fruit and vegetables on biomarkers of inflammation C-reactive protein, TNF-α, and in IL-6 in adolescents, decreasing F_2 isoprostane urinary levels [61]. Urinary neuroprostanes related to oxidative attack to the central nervous system (CNS) were reduced in elite athletes by the intake of lemon/aronia juice over 45 days, compared with a placebo rich in polyphenols, through the determination of F_2 isoprostanes and F_4 neuroprostanes like markers of oxidative stress in the CNS. Polyphenols reduce oxidative injury in neuronal membrane [62]. ROS are particularly dangerous in the CNS and are related to age and neurodegenerative diseases. In fact, another study revealed that supplementation with polyphenols decreased the effects of oxidative attack in the CNS and improved neural communicability through the consumption of blueberries and Concord grape [42].

Moreover, polyphenols reduced oxidative attack with moderate consumption of red wine, due to their ability to pass the blood–brain barrier, as demonstrated in an intervention measuring docosahexaenoic acid (DHA) peroxidation and production of F_2-dihomo-IsoPs and neuroprostanes as CNS oxidative injury markers of damage in the central nervous system, in that case hydroxitirosol [63]. However, due to the very low concentration of flavonoids detectable in the brain, it is unlikely that direct antioxidant action can entirely account for cognitive effects in vivo, according to other authors [64]. In fact, neurobiological effects are now believed to be mediated by actions involving the ability to protect vulnerable neurons, enhance neuronal function, and stimulate regeneration via interaction with neuronal intracellular signaling pathways involved in neuronal survival and differentiation, long-term potentiation, and memory [65].

3. Cognition

3.1. Cognitive Functions

Various cognitive functions, recognized by validated tests of the European Agency for Food Safety (EFSA) [66,67] are as follows:

- Sustained attention: the ability to maintain attention on a stimulus or a task for a long period.

- Selective attention: the ability to focus the mind on a task or a specific stimulus.
- Immediate memory: the ability to maintain a small amount of information in a short time period.
- Working memory: the set of processes that allow the storage and management of information for the performance of complex cognitive tasks such as language, reading, and mathematics.

3.2. Cognitive Impairment

Cognitive impairment is normally associated with advanced age or with degenerative diseases affecting cognition such as Alzheimer's disease or dementia [44,48,56,68–70]. The World Health Organization (WHO) estimates a new case of dementia every 4 seconds, which reveals the urgency of the global health crisis associated with cognitive impairment.

The global cost in 2010 in United States was estimated to be $604 billion. This figure equated to around 1% of the aggregated world gross domestic product (GDP) [71]. It is estimated that 10.5 million people in Europe have dementia, accounting for over 22% of the total number of people with dementia worldwide. The total societal costs of dementia in Europe in 2015 were estimated at $301 billion [72]. These diseases are growing in young people, and it has been proven that the conditions can be improved with the consumption of polyphenols through the diet. In fact, as discussed above, polyphenol consumption has been reported in 23 developed countries to decrease the rates of dementia [73], depression [74], and Alzheimer's disease [75].

For this reason, polyphenols can be considered potential compounds for the treatment and prevention of cognitive impairment [76,77], leading to improvement of the number and quality of synaptic connections in key brain regions [78], especially in the prevention of these pathologies due to the lack of effective symptomatic treatments [79].

In the case of the young population, lifestyle and nutritional habits become extremely important. This population has high basal metabolism requirements. Cell renovation and neuronal plasticity are higher in young people, so their nutritional requirements are more demanding, especially for preventing the loss of cognitive function. The adherence to the Mediterranean diet as a healthy dietary pattern results in the diminution of dementia risk in these young people when they reach an advanced age [80], and socio-economic factors do not affect the adherence to the Mediterranean diet, as shown in this study in Italy [81]. Polyphenol-rich products are beneficial for healthy brain function, particularly in early age, providing acute cognitive benefits. Benefits related to flavonoids and executive function in children have been showed with a recent publication describing positive results in a modified attention network task (MANT) [82,83], although the same author obtained contradictory results three years earlier in a similar research study [84].

On the other hand, related to executive function, other authors have associated the consumption of flavonoids in the young population with decreased risk of developing depression, linking flavonoid interventions with an improved positive mood [85]. This same relationship has also been evidenced by other authors [86,87], especially when it comes to the intake of phenolic acid, flavanones, and anthocyanins. If we talk about individual compounds, we have to especially consider quercetin and naringenin. Recently, a positive effect on the decrease of oxidative stress markers in a dietary pattern with high intake of fruits and vegetables and reduced meat consumption has also been reported [7,88,89].

Adult neurogenesis is the process associated with neural plasticity and the maintenance of normal function in the CNS, and therefore cognitive function and repair of damaged brain cells by aging. Diets enriched with polyphenols have been shown to induce neurogenesis in adult brains in the hippocampal region [90]. Neurogenesis is the process by which brain cells differentiate and proliferate into new neurons, and is a crucial factor in preserving cognitive function and repairing damaged brain cells affected by aging and brain disorders. There is a lack of scientific studies relating to dietary patterns and cognition [35,56] or acute cognitive benefits in the young population [91]. However, different unhealthy diets and lifestyle patterns have been associated with stress in adolescents [92]. In this sense, in Western societies, the trend to prevent or reverse childhood obesity problems by introducing morning and afternoon snacks may also ensure that young people have sufficient energy to complete daily cognitive tasks [93]. It is known that cognitive function declines with age [94], and that this is related to the consumption of antioxidants [56,95].

The improvement of cognition through the consumption of fruit and vegetables in subjects in whom this cognition has decreased due to age shows a greater relationship in the case of flavonoids, especially in anthocyanins and flavones [96]. Additionally, diabetic patients suffer greater oxidative stress and reduced antioxidant activity in the brain, and therefore show a greater cognitive impairment. This hypothesis was supported in a recent study with mice fed with grape and blueberry extracts, reporting possible mechanisms of action which may be associated with modulation of brain plasticity [97,98].

Polyphenols are safe treatment molecules for neurodegenerative diseases, presenting an alternative to drugs without any side effects of their therapeutic consumption. Numerous researchers have classified the beneficial effects of healthy food patterns, concluding that fruits and vegetables have immense potential as therapeutic food components [99].

The improvement of cognitive function has been related to nutrition and lifestyle [97], showing that neuronal damage caused by aging can be effectively countered with the supplementation of antioxidants such as polyphenols [38,100]. This fact has been recently reported in two interventional studies related to blueberry, and grape extracts [98,101]. Moreover, a recent pilot study showed a direct relationship between the consumption of blackcurrant polyphenols and the improvement of cognitive processes associated with attention [102].

4. Processes and Measuring Substances in Cognition Functions

4.1. Proteins CREB and BDNF (Brain-Derived Neurotrophic Factor)

The hippocampus is the major regulator of memory and learning processes (especially in long-term memory). Polyphenols are involved in neuronal plasticity through the protein CREB (Camp Response Element Binding), the protein kinase, and the activation of receptors in synapses, [24,103,104] modulating pathways of signaling and transcription factors. The phosphorylation of these kinases results in the modulation of synaptic efficacy [105,106]. This has been shown in tissue samples of the hippocampus in mice and the analysis of protein CREB [107]. The NF-kB factor is not only involved in inflammation processes, but in learning and the synaptic activity relationship with protein CREB, which is related to memory [108]. Others authors have reported the mechanism of action in the hippocampus and the receptors involved in the regulation. In neurological diseases and oxidative stress, receptors NMDA and GABA are excessively activated compared to normal situations, increasing synaptic plasticity in memory and teaching [109]. GABA is a neurotransmitter directly related to neurological disorders. For this reason, polyphenols have been acknowledged as potential chemical regulators providing an alternative to other pharmacological treatments. They may have a range of effects including relief of anxiety, improvement in cognition, and acting as neuroprotectants and as sedatives [110]. Recently, others authors have reported that treatment with flavonoids minimizes the increase of cytokines after stroke via the GABA receptor [111].

On the other hand, other factor in relation to synaptic plasticity and cognitive function, is Brain-Derived Neurotrophic Factor (BDNF). BDNF is a neurotrophin related to the maintenance, survival, growth, and differentiation of neurons [112]. In fact, BDNF is so essential to the brain's cognitive processes that deletion of the BDNF gene was found to weaken memory retention and inhibit long-term potentiation [113]. Recent findings have shown neuroprotective effects of quercetin due to the increment of BDNF mRNA [114,115]. Moreover, there seems to be conflicting results about whether this neurotrophin is genetically influenced by neurodegenerative diseases [116]. Indeed, to date there has been no study to clarify whether the reduction of BDNF levels in neurodegenerative diseases is a downstream consequence of the disease process or part of a specific disease mechanism.

In a recent study with rats, it was proven that the level of BDNF was reduced in hippocampus by the administration of carob polyphenols in induced chronic stress [117], learning, and memory [118].

4.2. Cerebral Blood Flow

Flavonoids, especially anthocyanins, have a positive effect on the brain cells associated with memory and neuronal function, mainly due to the increase of cerebral blood flow [119–121]. This increase in the blood flow to the brain causes an improvement in synaptic plasticity, leading to the improvement of cognitive capacities, shown especially in animal studies and with flavonoid-enriched diets containing grape, pomegranate, strawberry, blueberry, cocoa and pure flavonoids, such as quercetin [122–126]. Authors have reported on not only the increase of polyphenol intake, but also physical activity and intake of alcohol and caffeine. Decreases in cerebral blood flow are induced by commonly consumed amounts of caffeine, while alcohol significantly increases cerebral perfusion in a dose-dependent way [127]. Additionally, lifestyle factors, including dietary composition and physical exercise, can significantly increase CBF, thereby improving cognitive performance.

Included in flavonoids, catechins participate in the signaling cascades of the kinase lipid transcription factors like the CREB protein, as well as changes in the vascular system and the brain [56]. In particular, the connections between neurons through the ERK and AKT signaling pathways are increased, promoting neurotrophies such as BDNF [128].

Moreover, in a recent study on mice regarding oxidative stress and neurodegenerative diseases, improvement of cognitive functions after administration of quercetin over 13 weeks was observed. These results showed an inverse relationship between oxidative stress and neurodegenerative diseases, judging by the reduction of BDNF and CREB and modulation of AKT. Quercetin improved cognitive function as well as increasing BDNF, CREB, and AKT when was administrated over 13 weeks [129]. In the same way, other authors have described a positive protective effect against oxidative-stress-triggered cognitive impairment related to the intake of green tea polyphenols [130]. Finally, epigallocatechin gallate, epicatechin, anthocyanins, and pelargonidins are flavonoids able to cross the blood–brain barrier and therefore exert great effect on cognitive function. This cognitive improvement seems to be regulated by the interaction with ERK and AKT pathways, leading to the modulation of CREB factor and regulation of CREB gene expression [104,131].

Furthermore, the epigallocatechin gallate is able to exert neuroprotection against corticosterone-induced neuron damage by restoring the ERK and PI3K/AKT signaling pathways [132]. The relationship not only between these factors and polyphenols has been demonstrated recently, but also between obesity and insulin resistance [133].

The increase in brain blood flow has also been revealed not only by cognitive test, but also by magnetic resonance in healthy young adults and by using near-infrared spectroscopy (NIRS) in a study with epigallocatechin gallate [29,120,134,135]. The mechanism of action associated with vasodilation is the liberation of nitric oxide through nitric oxide synthase with an improvement in endothelial function [136]. It was consistent with a study on consumption of a cocoa beverage rich in flavonols [137], as well as other authors' work [138]. A major compound related to these effects is epicatechin, the main flavanol present in cocoa, as evidenced in the study with the Kuna Indians of Panama [139]. That study on cocoa flavonoids supported the idea that flavanol-induced acute changes in brain perfusion could underpin the immediate cognitive-enhancing effects [140]. Accordingly, other authors have reported a direct relationship between the consumption of flavonoids and vasodilation due to nitric oxide synthase activation [137].

Thus, flavonoids are considered important in cognitive function, influencing signaling pathways that are involved in processing of memory. However, the mechanisms of action have not yet been clarified [136]. In this sense, other authors have described similar effects related to resveratrol, assessing the effects on cognitive performance and cerebral brain blood and modulating concentration of hemoglobin and cerebral blood flow, checked with cognitive tasks, NIRS, and determination of hemoglobin and deoxyhemoglobin [141].

Flavonoids act within the cell signaling chain of protein kinases, namely PI3 kinase/AKT. These cascades stimulate gene expression and protein synthesis to maintain long-term potentiation, establish long-term memory [106], modulate transcription factors that participate in signal transduction through protein kinase inhibition, and promote the expression of brain-derived neurotrophic factor (BDNF) that is critical for neurogenesis [142]. In fact, chronic administration of polyphenols has this effect on hippocampal memory by phosphorylation of ERK, CREB, and the expression of BDNF [143].

As some authors have described, the hormone 17β-estradiol has a critical role in the ERK signaling pathway, which had not previously been identified. Therefore, neuron-derived estradiol functions may be a novel neuromodulator in the brain, related to the control of synaptic plasticity and cognitive function [144].

4.3. Memory

The protective effects of polyphenols regarding cognition and memory improvement have been highlighted in the case of green tea, mainly associating the epigallocatechin gallate with effects in working memory and attention [145]; through metabolites able to reach the brain parenchyma [146,147]; cocoa polyphenols, as shown during a study of psychological stress in rats [44]; and other polyphenols such as quercetin, which have been shown to be useful in the prevention and treatment of degenerative diseases in humans related to memory [148,149]. This effect appears to be defined by a significant decrease in biomarkers of stress and inflammation, such as dopamine and norepinephrine, cortisol, IL-6, and superoxide dismutase. Quercetin improves learning and memory-impairment-attenuating scopolamine-induced cholinesterase activity and the BDNF level in the prefrontal cortex and hippocampus [150].

There seems to be an improvement of cognitive functions in interventional studies regarding consumption of polyphenol-rich foods, especially flavonoids. These effects appear to be obvious in very young or elderly populations in whom the neural growth of hippocampus or neuronal connectivity can be improved [121]. In particular, the region of the hippocampus which is implicated in memory, especially in older adults with neurodegenerative diseases, is the dentate gyrus [149,151]. This fact has been demonstrated with a resveratrol-based nutraceutical in young adults, subjecting them to demanding cognitive tasks and relating them to brain blood flow through the modulation of nitric oxide [135,152] as well as improving endothelial function [153]. In the same way, flavonoid consumption produced an improvement of the adult proliferation rate in hippocampus in adult hippocampal neurogenesis, directly linked to cognition, proving a potent antioxidant capacity [154].

In a prospective study in French adults, memory, executive function, and language were evaluated. It was not possible to determine a clear and evident association between improvement of these functions and the consumption of fruits or vegetables. However, that relationship did occur with the improvement of cognitive impairment [37]. Likewise, the consumption of multivitamins and their composition has been linked to improvements of cognitive performance in some tasks associated with different types of memory [155].

Despite the above, randomized scientific studies reporting on polyphenols as potential agents on memory and learning seem to be more focused on elderly people (especially related to neurodegenerative diseases) than on young people [104,156] especially in Alzheimer's disease [157,158].

5. Conclusions and Future Perspectives

Large amounts of previous literature have demonstrated the positive effect of polyphenols related to neuroprotection and antioxidant and anti-inflammatory capacity. As regards memory and cognitive functions, other authors have reported interventional studies showing their improvement by increased brain blood flow and BDNF- or CREB-protein-related pathways.

In this context, an adequate intake of fruits and vegetables is necessary for the maintenance of normal cognitive functions. The population with cognitive deterioration has been increasing around the world since the second half of the 20th century. A further decline in cognition might be preventable in the early stages of cognitive impairment by regular intake of polyphenols present in foods. The mechanisms of flavonoids have been shown through the inhibition of free radicals and modulation of signaling pathways that are implicated in cognitive and neuroprotective functions [52,159].

In order to reach an adequate consumption of polyphenols, nutraceuticals are food that can provide an easy and simple way to consume the recommended daily allowance of polyphenols, although more knowledge is necessary about their actual effects on human health bioavailability, pharmacokinetics, and mechanisms of action [160,161]. The nutraceutical industry is becoming the new food industry due to the incorporation of these bioactive substances into our daily lives [162,163].

Additionally, there is a lack of effective pharmaceutical treatments for age-related cognitive decline [164], but an increase of polyphenol-rich nutritional supplements and botanical extracts could be an alternative [76,165].

Related to cognitive impairment, these supplements and vegetarian foods are good options for the improvement of mood and cognitive function in the healthy population. The importance of food supplements and nutraceuticals includes economic factors. A food supplement as cognitive deterioration treatment that would reduce the deterioration by 1% per year could reduce the monetary cost of long-term treatment of patients with cognitive dysfunctions. Therefore, fruits and vegetables could be main foods for the improvement of cognitive functions [96].

Cognitive function has been associated with the consumption of fruit and vegetables in patients with degenerative diseases However, there is a lack of information about interventional studies in humans under clinical conditions [54,166] in healthy, young populations [167], especially related to dietary patterns and cognition [35], reporting neurological benefits associated with individual intake of fruit and vegetables or their synergic effect [4]. There have also not been enough intervention studies in humans about the gut–microbiota–brain axis, which has been proven to be one of the lines by which diet may improve the development and healthy functions of the brain [33].

Therefore, more studies in young and healthy populations seems to be relevant not only for these populations, but also to find out the consequences of chronic consumption of fruits and vegetables on cognitive functions, thus implementing primary prevention actions in young people to minimize or slow down the cognitive impairment caused by age.

Funding: This research received no external funding.

Conflicts of Interest: The authors declare no conflict of interest.

References

1. World Health Organization. *Diet, Nutrition and the Prevention of Chronic Diseases: Report of a Joint WH/FAO Expert Consultation*; World Health Organization: Geneva, Switzerland, 2003.

2. World Health Organization. *European Action Plan for Food and Nutrition Policy 2007–2012*; WHO Technical Report Series; World Health Organization: Geneva, Switzerland, 2008.

3. Aranceta, J.; Pérez-Rodrigo, C. Recommended dietary reference intakes, nutritional goals and dietary guidelines for fat and fatty acids: A systematic review. *Br. J. Nutr.* **2012**, *107*, S8–S22. [CrossRef] [PubMed]

4. Miller, M.G.; Thangthaeng, N.; Poulose, S.M.; Shukitt-Hale, B. Role of fruits, nuts, and vegetables in maintaining cognitive health. *Exp. Gerontol.* **2017**, *94*, 24–28. [CrossRef] [PubMed]

5. Ma, W.; Hagan, K.A.; Heianza, Y.; Sun, Q.; Rimm, E.B.; Qi, L. Adult height, dietary patterns, and healthy aging. *Am. J. Clin. Nutr.* **2017**, *106*, 589–596. [CrossRef] [PubMed]

6. Saura-Calixto, F.; Goni, I. Definition of the Mediterranean diet based on bioactive compounds. *Crit. Rev. Food Sci. Nutr.* **2009**, *49*, 145–152. [CrossRef]

7. Meyer, K.A.; Sijtsma, F.P.; Nettleton, J.A.; Steffen, L.M.; Van Horn, L.; Shikany, J.M.; Gross, M.D.; Mursu, J.; Traber, M.G.; Jacobs, D.R., Jr. Dietary patterns are associated with plasma F2-isoprostanes in an observational cohort study of adults. *Free Radicals Biol. Med.* **2013**, *57*, 201–209. [CrossRef]

8. Sofi, F.; Macchi, C.; Abbate, R.; Gensini, G.F.; Casini, A. Mediterranean diet and health status: An updated meta-analysis and a proposal for a literature-based adherence score. *Public Health Nutr.* **2014**, *17*, 2769–2782. [CrossRef]

9.   Pérez-Jiménez, J.; Díaz-Rubio, M.E.; Saura-Calixto, F. Contribution of macromolecular antioxidants to dietary antioxidant capacity: A Study in the Spanish Mediterranean Diet. *Plant Foods Hum. Nutr.* **2015**, *70*, 365–370. [CrossRef]

10.   Yu, F.-N.; Hu, N.-Q.; Huang, X.-L.; Shi, Y.-X.; Zhao, H.-Z.; Cheng, H.-Y. Dietary patterns derived by factor analysis are associated with cognitive function among a middle-aged and elder Chinese population. *Psychiatry Res.* **2018**, *269*, 640–645. [CrossRef]

11.   Lamprecht, M.; Obermayer, G.; Steinbauer, K.; Cvirn, G.; Hofmann, L.; Ledinski, G.; Greilberger, J.F.; Hallstroem, S. Supplementation with a juice powder concentrate and exercise decrease oxidation and inflammation, and improve the microcirculation in obese women: Randomised controlled trial data. *Br. J. Nutr.* **2013**, *110*, 1685–1695. [CrossRef]

12.   Dreiseitel, A.; Schreier, P.; Oehme, A.; Locher, S.; Rogler, G.; Piberger, H.; Hajak, G.; Sand, P.G. Inhibition of proteasome activity by anthocyanins and anthocyanidins. *Biochem. Biophys. Res. Commun.* **2008**, *372*, 57–61. [CrossRef]

13.   Traustadóttir, T.; Davies, S.S.; Stock, A.A.; Su, Y.; Heward, C.B.; Roberts, L.J.; Harman, S.M. Tart cherry juice decreases oxidative stress in healthy older men and women. *J. Nutr.* **2009**, *139*, 1896–1900. [CrossRef]

14.   Tressera-Rimbau, A.; Arranz, S.; Eder, M.; Vallverdú-Queralt, A. Dietary polyphenols in the prevention of stroke. *Oxid. Med. Cell. Longev.* **2017**, *2017*. [CrossRef] [PubMed]

15.   Dragsted, L.O.; Krath, B.; Ravn-Haren, G.; Vogel, U.B.; Vinggaard, A.M.; Jensen, P.B.; Loft, S.; Rasmussen, S.E.; Sandstrom, B.; Pedersen, A. Biological effects of fruit and vegetables. *Proc. Nutr. Soc.* **2006**, *65*, 61–67. [CrossRef] [PubMed]

16.   Bellone, J.A.; Murray, J.R.; Jorge, P.; Fogel, T.G.; Kim, M.; Wallace, D.R.; Hartman, R.E. Pomegranate supplementation improves cognitive and functional recovery following ischemic stroke: A randomized trial. *Nutr. Neurosci.* **2019**, *22*, 738–743. [CrossRef] [PubMed]

17.   Angelino, D.; Godos, J.; Ghelfi, F.; Tieri, M.; Titta, L.; Lafranconi, A.; Marventano, S.; Alonzo, E.; Gambera, A.; Sciacca, S. Fruit and vegetable consumption and health outcomes: An umbrella review of observational studies. *Int. J. Food Sci. Nutr.* **2019**, *70*, 652–667. [CrossRef] [PubMed]

18.   Zhang, Y.; Zhang, X.; Cui, L.; Chen, R.; Zhang, C.; Li, Y.; He, T.; Zhu, X.; Shen, Z.; Dong, L. Salvianolic Acids for Injection (SAFI) promotes functional recovery and neurogenesis via sonic hedgehog pathway after stroke in mice. *Neurochem. Int.* **2017**, *110*, 38–48. [CrossRef] [PubMed]

19.   Xu, H.; Luo, J.; Huang, J.; Wen, Q. Flavonoids intake and risk of type 2 diabetes mellitus: A meta-analysis of prospective cohort studies. *Medicine* **2018**, *97*. [CrossRef]

20.   Grosso, G.; Godos, J.; Lamuela-Raventos, R.; Ray, S.; Micek, A.; Pajak, A.; Sciacca, S.; D'orazio, N.; Del Rio, D.; Galvano, F. A comprehensive meta-analysis on dietary flavonoid and lignan intake and cancer risk: Level of evidence and limitations. *Mol. Nutr. Food Res.* **2017**, *61*. [CrossRef]

21.   Grosso, G.; Micek, A.; Godos, J.; Pajak, A.; Sciacca, S.; Galvano, F.; Giovannucci, E.L. Dietary flavonoid and lignan intake and mortality in prospective cohort studies: Systematic review and dose-response meta-analysis. *Am. J. Epidemiol.* **2017**, *185*, 1304–1316. [CrossRef]

22.   Péter, S.; Holguin, F.; Wood, L.G.; Clougherty, J.E.; Raederstorff, D.; Antal, M.; Weber, P.; Eggersdorfer, M. Nutritional solutions to reduce risks of negative health impacts of air pollution. *Nutrients* **2015**, *7*, 10398–10416. [CrossRef]

23.   Nile, S.H.; Park, S.W. Edible berries: Bioactive components and their effect on human health. *Nutrition* **2014**, *30*, 134–144. [CrossRef] [PubMed]

24.   Miller, M.G.; Shukitt-Hale, B. Berry fruit enhances beneficial signaling in the brain. *J. Agric. Food Chem.* **2012**, *60*, 5709–5715. [CrossRef] [PubMed]

25.   Guo, C.; Wei, J.; Yang, J.; Xu, J.; Pang, W.; Jiang, Y. Pomegranate juice is potentially better than apple juice in improving antioxidant function in elderly subjects. *Nutr. Res.* **2008**, *28*, 72–77. [CrossRef] [PubMed]

26.   Battino, M.; Beekwilder, J.; Denoyes-Rothan, B.; Laimer, M.; McDougall, G.J.; Mezzetti, B. Bioactive compounds in berries relevant to human health. *Nutr. Rev.* **2009**, *67*, S145–S150. [CrossRef] [PubMed]

27.   Skrovankova, S.; Sumczynski, D.; Mlcek, J.; Jurikova, T.; Sochor, J. Bioactive compounds and antioxidant activity in different types of berries. *Int. J. Mol. Sci.* **2015**, *16*, 24673–24706. [CrossRef]

28.   Hollman, P.C.; Cassidy, A.; Comte, B.; Heinonen, M.; Richelle, M.; Richling, E.; Serafini, M.; Scalbert, A.; Sies, H.; Vidry, S. The Biological Relevance of Direct Antioxidant Effects of Polyphenols for Cardiovascular Health in Humans is Not Established. *J. Nutr.* **2011**, *141*, 989S–1009S. [CrossRef]

29. Lamport, D.J.; Lawton, C.L.; Merat, N.; Jamson, H.; Myrissa, K.; Hofman, D.; Chadwick, H.K.; Quadt, F.; Wightman, J.D.; Dye, L. Concord grape juice, cognitive function, and driving performance: A 12-wk, placebo-controlled, randomized crossover trial in mothers of preteen children. *Am. J. Clin. Nutr.* **2016**, *103*, 775–783. [CrossRef]

30. Verhagen, H.; Coolen, S.; Duchateau, G.; Hamer, M.; Kyle, J.; Rechner, A. Assessment of the efficacy of functional food ingredients—introducing the concept "kinetics of biomarkers". *Mutat. Res. Fundam. Mol. Mech. Mutagen.* **2004**, *551*, 65–78. [CrossRef]

31. Del Rio, D.; Rodriguez-Mateos, A.; Spencer, J.P.; Tognolini, M.; Borges, G.; Crozier, A. Dietary (poly) phenolics in human health: Structures, bioavailability, and evidence of protective effects against chronic diseases. *Antioxid. Redox Signal.* **2013**, *18*, 1818–1892. [CrossRef]

32. Pereira-Caro, G.; Borges, G.; Ky, I.; Ribas, A.; Calani, L.; Del Rio, D.; Clifford, M.N.; Roberts, S.A.; Crozier, A. In vitro colonic catabolism of orange juice (poly) phenols. *Mol. Nutr. Food Res.* **2015**, *59*, 465–475. [CrossRef]

33. Ceppa, F.; Mancini, A.; Tuohy, K. Current evidence linking diet to gut microbiota and brain development and function. *Int. J. Food Sci. Nutr.* **2019**, *70*, 1–19. [CrossRef] [PubMed]

34. Welch, R.W.; Antoine, J.-M.; Berta, J.-L.; Bub, A.; de Vries, J.; Guarner, F.; Hasselwander, O.; Hendriks, H.; Jäkel, M.; Koletzko, B.V. Guidelines for the design, conduct and reporting of human intervention studies to evaluate the health benefits of foods. *Br. J. Nutr.* **2011**, *106*, S3–S15. [CrossRef] [PubMed]

35. Zhu, N.; Jacobs, D.R.; Meyer, K.; He, K.; Launer, L.; Reis, J.; Yaffe, K.; Sidney, S.; Whitmer, R.; Steffen, L. Cognitive function in a middle aged cohort is related to higher quality dietary pattern 5 and 25 years earlier: The CARDIA study. *J. Nutr. Health Aging* **2015**, *19*, 33–38. [CrossRef] [PubMed]

36. Sabia, S.; Nabi, H.; Kivimaki, M.; Shipley, M.J.; Marmot, M.G.; Singh-Manoux, A. Health behaviors from early to late midlife as predictors of cognitive function: The Whitehall II study. *Am. J. Epidemiol.* **2009**, *170*, 428–437. [CrossRef] [PubMed]

37. Péneau, S.; Galan, P.; Jeandel, C.; Ferry, M.; Andreeva, V.; Hercberg, S.; Kesse-Guyot, E.; SU.VI.MAX 2 Research Group. Fruit and vegetable intake and cognitive function in the SU. VI. MAX 2 prospective study. *Am. J. Clin. Nutr.* **2011**, *94*, 1295–1303. [CrossRef] [PubMed]

38. Joseph, J.A.; Shukitt-Hale, B.; Casadesus, G. Reversing the deleterious effects of aging on neuronal communication and behavior: Beneficial properties of fruit polyphenolic compounds. *Am. J. Clin. Nutr.* **2005**, *81*, 313S–316S. [CrossRef] [PubMed]

39. Tsao, R. Chemistry and biochemistry of dietary polyphenols. *Nutrients* **2010**, *2*, 1231–1246. [CrossRef]

40. Panche, A.; Diwan, A.; Chandra, S. Flavonoids: An overview. *J. Nutr. Sci.* **2016**, *5*. [CrossRef]

41. Williams, C.A.; Grayer, R.J. Anthocyanins and other flavonoids. *Nat. Prod. Rep.* **2004**, *21*, 539–573. [CrossRef]

42. Joseph, J.A.; Shukitt-Hale, B.; Willis, L.M. Grape juice, berries, and walnuts affect brain aging and behavior. *J. Nutr.* **2009**, *139*, 1813S–1817S. [CrossRef]

43. Yamada, K.; Tachibana, H. Recent topics in anti-oxidative factors. *BioFactors* **2000**, *13*, 167–172. [CrossRef] [PubMed]

44. Chen, W.-Q.; Zhao, X.-L.; Hou, Y.; Li, S.-T.; Hong, Y.; Wang, D.-L.; Cheng, Y.-Y. Protective effects of green tea polyphenols on cognitive impairments induced by psychological stress in rats. *Behav. Brain Res.* **2009**, *202*, 71–76. [CrossRef] [PubMed]

45. Rauf, A.; Imran, M.; Suleria, H.A.R.; Ahmad, B.; Peters, D.G.; Mubarak, M.S. A comprehensive review of the health perspectives of resveratrol. *Food Funct.* **2017**, *8*, 4284–4305. [CrossRef] [PubMed]

46. Straniero, S.; Cavalini, G.; Donati, A.; Bergamini, E. Resveratrol requires red wine polyphenols for optimum antioxidant activity. *J. Nutr. Health Aging* **2016**, *20*, 540–545.

47. Khan, I.; Yousif, A.M.; Johnson, S.K.; Gamlath, S. Acute effect of sorghum flour-containing pasta on plasma total polyphenols, antioxidant capacity and oxidative stress markers in healthy subjects: A randomised controlled trial. *Clin. Nutr.* **2015**, *34*, 415–421. [CrossRef]

48. Weisel, T.; Baum, M.; Eisenbrand, G.; Dietrich, H.; Will, F.; Stockis, J.P.; Kulling, S.; Rüfer, C.; Johannes, C.; Janzowski, C. An anthocyanin/polyphenolic-rich fruit juice reduces oxidative DNA damage and increases glutathione level in healthy probands. *Biotechnol. J.* **2006**, *1*, 388–397. [CrossRef]

49. Rahman, M.M.; Ichiyanagi, T.; Komiyama, T.; Sato, S.; Konishi, T. Effects of anthocyanins on psychological stress-induced oxidative stress and neurotransmitter status. *J. Agric. Food Chem.* **2008**, *56*, 7545–7550. [CrossRef]

50. Dulebohn, R.V.; Yi, W.; Srivastava, A.; Akoh, C.C.; Krewer, G.; Fischer, J.G. Effects of blueberry (*Vaccinium ashei*) on DNA damage, lipid peroxidation, and phase II enzyme activities in rats. *J. Agric. Food Chem.* **2008**, *56*, 11700–11706. [CrossRef]

51. Martínez-Tomás, R.; Larqué, E.; González-Silvera, D.; Sánchez-Campillo, M.; Burgos, M.I.; Wellner, A.; Parra, S.; Bialek, L.; Alminger, M.; Pérez-Llamas, F. Effect of the consumption of a fruit and vegetable soup with high in vitro carotenoid bioaccessibility on serum carotenoid concentrations and markers of oxidative stress in young men. *Eur. J. Nutr.* **2012**, *51*, 231–239. [CrossRef]

52. Castelli, V.; Grassi, D.; Bocale, R.; d'Angelo, M.; Antonosante, A.; Cimini, A.; Ferri, C.; Desideri, G. Diet and Brain Health: Which Role for Polyphenols? *Curr. Pharm. Des.* **2018**, *24*, 227–238. [CrossRef]

53. Calder, P.C.; Albers, R.; Antoine, J.-M.; Blum, S.; Bourdet-Sicard, R.; Ferns, G.; Folkerts, G.; Friedmann, P.; Frost, G.; Guarner, F. Inflammatory disease processes and interactions with nutrition. *Br. J. Nutr.* **2009**, *101*, 1–45. [CrossRef] [PubMed]

54. Gildawie, K.R.; Galli, R.L.; Shukitt-Hale, B.; Carey, A.N. Protective Effects of Foods Containing Flavonoids on Age-Related Cognitive Decline. *Curr. Nutr. Rep.* **2018**, *7*, 39–48. [CrossRef] [PubMed]

55. Windham, B.G.; Simpson, B.N.; Lirette, S.; Bridges, J.; Bielak, L.; Peyser, P.A.; Kullo, I.; Turner, S.; Griswold, M.E.; Mosley, T.H. Associations between inflammation and cognitive function in African Americans and European Americans. *J. Am. Geriatr. Soc.* **2014**, *62*, 2303–2310. [CrossRef] [PubMed]

56. Rodriguez-Mateos, A.; Vauzour, D.; Krueger, C.G.; Shanmuganayagam, D.; Reed, J.; Calani, L.; Mena, P.; Del Rio, D.; Crozier, A. Bioavailability, bioactivity and impact on health of dietary flavonoids and related compounds: An update. *Arch. Toxicol.* **2014**, *88*, 1803–1853. [CrossRef] [PubMed]

57. Huang, X.; Pan, Q.; Mao, Z.; Wang, P.; Zhang, R.; Ma, X.; Chen, J.; You, H. Kaempferol inhibits interleukin-1β stimulated matrix metalloproteinases by suppressing the MAPK-associated ERK and P38 signaling pathways. *Mol. Med. Rep.* **2018**, *18*, 2697–2704. [CrossRef] [PubMed]

58. Wu, P.-S.; Ding, H.-Y.; Yen, J.-H.; Chen, S.-F.; Lee, K.-H.; Wu, M.-J. Anti-inflammatory activity of 8-hydroxydaidzein in LPS-stimulated BV2 microglial cells via activation of Nrf2-antioxidant and attenuation of Akt/NF-κB-inflammatory signaling pathways, as well as inhibition of COX-2 activity. *J. Agric. Food Chem.* **2018**, *66*, 5790–5801. [CrossRef]

59. Grabs, V.; Nieman, D.C.; Haller, B.; Halle, M.; Scherr, J. The effects of oral hydrolytic enzymes and flavonoids on inflammatory markers and coagulation after marathon running: Study protocol for a randomized, double-blind, placebo-controlled trial. *Sports Sci. Med. Rehab.* **2014**, *6*, 8. [CrossRef]

60. Hermsdorff, H.H.M.; Barbosa, K.B.; Volp, A.C.P.; Puchau, B.; Bressan, J.; Zulet, M.A.; Martínez, J.A. Vitamin C and fibre consumption from fruits and vegetables improves oxidative stress markers in healthy young adults. *Br. J. Nutr.* **2012**, *107*, 1119–1127. [CrossRef]

61. Holt, E.M.; Steffen, L.M.; Moran, A.; Basu, S.; Steinberger, J.; Ross, J.A.; Hong, C.-P.; Sinaiko, A.R. Fruit and vegetable consumption and its relation to markers of inflammation and oxidative stress in adolescents. *J. Am. Diet. Assoc.* **2009**, *109*, 414–421. [CrossRef]

62. García-Flores, L.A.; Medina, S.; Oger, C.; Galano, J.-M.; Durand, T.; Cejuela, R.; Martínez-Sanz, J.M.; Ferreres, F.; Gil-Izquierdo, Á. Lipidomic approach in young adult triathletes: Effect of supplementation with a polyphenols-rich juice on neuroprostane and F 2-dihomo-isoprostane markers. *Food Funct.* **2016**, *7*, 4343–4355. [CrossRef]

63. Marhuenda, J.; Medina, S.; Martínez-Hernández, P.; Arina, S.; Zafrilla, P.; Mulero, J.; Oger, C.; Galano, J.-M.; Durand, T.; Ferreres, F. Melatonin and hydroxytyrosol protect against oxidative stress related to the central nervous system after the ingestion of three types of wine by healthy volunteers. *Food Funct.* **2017**, *8*, 64–74. [CrossRef] [PubMed]

64. Socci, V.; Tempesta, D.; Desideri, G.; De Gennaro, L.; Ferrara, M. Enhancing Human cognition with cocoa flavonoids. *Front. Nutr.* **2017**, *4*, 19. [CrossRef] [PubMed]

65. Spencer, J.P. Food for thought: The role of dietary flavonoids in enhancing human memory, learning and neuro-cognitive performance: Symposium on 'diet and mental health'. *Proc. Nutr. Soc.* **2008**, *67*, 238–252. [CrossRef] [PubMed]

66. Gilsenan, M. Nutrition & health claims in the European Union: A regulatory overview. *Trends Food Sci. Technol.* **2011**, *22*, 536–542.

67. De Jager, C.A.; Dye, L.; de Bruin, E.A.; Butler, L.; Fletcher, J.; Lamport, D.J.; Latulippe, M.E.; Spencer, J.P.; Wesnes, K. Criteria for validation and selection of cognitive tests for investigating the effects of foods and nutrients. *Nutr. Rev.* **2014**, *72*, 162–179. [CrossRef] [PubMed]

68. Gu, Y.; Scarmeas, N. Dietary patterns in Alzheimer's disease and cognitive aging. *Curr. Alzheimer Res.* **2011**, *8*, 510–519. [CrossRef] [PubMed]

69. Thapa, A.; Carroll, N.J. Dietary Modulation of Oxidative Stress in Alzheimer's Disease. *Int. J. Mol. Sci.* **2017**, *18*, 1583. [CrossRef]

70. Wimo, A.; Guerchet, M.; Ali, G.-C.; Wu, Y.-T.; Prina, A.M.; Winblad, B.; Jönsson, L.; Liu, Z.; Prince, M. The worldwide costs of dementia 2015 and comparisons with 2010. *Alzheimer's Dement.* **2017**, *13*, 1–7. [CrossRef]

71. Handels, R.L.; Sköldunger, A.; Bieber, A.; Edwards, R.T.; Gonçalves-Pereira, M.; Hopper, L.; Irving, K.; Jelley, H.; Kerpershoek, L.; Marques, M.J. Quality of life, care resource use, and costs of dementia in 8 European countries in a cross-sectional cohort of the actifcare study. *J. Alzheimer's Dis.* **2018**, *66*, 1027–1040. [CrossRef]

72. Beking, K.; Vieira, A. Flavonoid intake and disability-adjusted life years due to Alzheimer's and related dementias: A population-based study involving twenty-three developed countries. *Public Health Nutr.* **2010**, *13*, 1403–1409. [CrossRef]

73. Calapai, G.; Crupi, A.; Firenzuoli, F.; Inferrera, G.; Squadrito, F.; Parisi, A.; De Sarro, G.; Caputi, A. Serotonin, norepinephrine and dopamine involvement in the antidepressant action of hypericum perforatum. *Pharmacopsychiatry* **2001**, *34*, 45–49. [CrossRef] [PubMed]

74. Bui, T.T.; Nguyen, T.H. Natural product for the treatment of Alzheimer's disease. *J. Basic Clin. Physiol. Pharmacol.* **2017**, *28*, 413–423. [CrossRef] [PubMed]

75. Hussain, G.; Zhang, L.; Rasul, A.; Anwar, H.; Sohail, M.U.; Razzaq, A.; Aziz, N.; Shabbir, A.; Ali, M.; Sun, T. Role of Plant-Derived Flavonoids and Their Mechanism in Attenuation of Alzheimer's and Parkinson's Diseases: An Update of Recent Data. *Molecules* **2018**, *23*, 814. [CrossRef] [PubMed]

76. Scholey, A. Nutrients for neurocognition in health and disease: Measures, methodologies and mechanisms. *Proc. Nutr. Soc.* **2018**, *77*, 73–83. [CrossRef]

77. Moore, K.; Hughes, C.F.; Ward, M.; Hoey, L.; McNulty, H. Diet, nutrition and the ageing brain: Current evidence and new directions. *Proc. Nutr. Soc.* **2018**, *77*, 152–163. [CrossRef]

78. Williams, R.J.; Spencer, J.P. Flavonoids, cognition, and dementia: Actions, mechanisms, and potential therapeutic utility for Alzheimer disease. *Free Radicals Biol. Med.* **2012**, *52*, 35–45. [CrossRef]

79. Olivera-Pueyo, J.; Pelegrín-Valero, C. Dietary supplements for cognitive impairment. *Actas Esp. Psiquiatr.* **2017**, *45*, 37–47.

80. Feart, C.; Samieri, C.; Rondeau, V.; Amieva, H.; Portet, F.; Dartigues, J.-F.; Scarmeas, N.; Barberger-Gateau, P. Adherence to a Mediterranean diet, cognitive decline, and risk of dementia. *JAMA* **2009**, *302*, 638–648. [CrossRef]

81. Buonomo, E.; Moramarco, S.; Tappa, A.; Palmieri, S.; Di Michele, S.; Biondi, G.; Agosti, G.; Alessandroni, C.; Caredda, E.; Palombi, L. Access to health care, nutrition and dietary habits among school-age children living in socio-economic inequality contexts: Results from the "ForGood: Sport is Well-Being" programme. *Int. J. Food Sci. Nutr.* **2019**, 1–10. [CrossRef]

82. Whyte, A.R.; Schafer, G.; Williams, C.M. The effect of cognitive demand on performance of an executive function task following wild blueberry supplementation in 7 to 10 years old children. *Food Funct.* **2017**, *8*, 4129–4138. [CrossRef]

83. Barfoot, K.L.; May, G.; Lamport, D.J.; Ricketts, J.; Riddell, P.M.; Williams, C.M. The effects of acute wild blueberry supplementation on the cognition of 7–10-year-old schoolchildren. *Eur. J. Nutr.* **2018**, *58*, 2911–2920. [CrossRef] [PubMed]

84. Whyte, A.R.; Williams, C.M. Effects of a single dose of a flavonoid-rich blueberry drink on memory in 8 to 10 y old children. *Nutrition* **2015**, *31*, 531–534. [CrossRef] [PubMed]

85. Khalid, S.; Barfoot, K.; May, G.; Lamport, D.; Reynolds, S.; Williams, C. Effects of acute blueberry flavonoids on mood in children and young adults. *Nutrients* **2017**, *9*, 158. [CrossRef] [PubMed]

86. Chang, S.-C.; Cassidy, A.; Willett, W.C.; Rimm, E.B.; O'Reilly, E.J.; Okereke, O.I. Dietary flavonoid intake and risk of incident depression in midlife and older women. *Am. J. Clin. Nutr.* **2016**, *104*, 704–714. [CrossRef]

87. Godos, J.; Castellano, S.; Ray, S.; Grosso, G.; Galvano, F. Dietary polyphenol intake and depression: Results from the mediterranean healthy eating, lifestyle and aging (meal) study. *Molecules* **2018**, *23*, 999. [CrossRef]

88. Pistollato, F.; Battino, M. Role of plant-based diets in the prevention and regression of metabolic syndrome and neurodegenerative diseases. *Trends Food Sci. Technol.* **2014**, *40*, 62–81. [CrossRef]

89. Pistollato, F.; Sumalla Cano, S.; Elio, I.; Masias Vergara, M.; Giampieri, F.; Battino, M. Associations between sleep, cortisol regulation, and diet: Possible implications for the risk of Alzheimer disease. *Adv. Nutr.* **2016**, *7*, 679–689. [CrossRef]

90. Poulose, S.M.; Miller, M.G.; Scott, T.; Shukitt-Hale, B. Nutritional factors affecting adult neurogenesis and cognitive function. *Adv. Nutr.* **2017**, *8*, 804–811. [CrossRef]

91. Haskell-Ramsay, C.; Stuart, R.; Okello, E.; Watson, A. Cognitive and mood improvements following acute supplementation with purple grape juice in healthy young adults. *Eur. J. Nutr.* **2017**, *56*, 2621–2631. [CrossRef]

92. Weng, T.-T.; Hao, J.-H.; Qian, Q.-W.; Cao, H.; Fu, J.-L.; Sun, Y.; Huang, L.; Tao, F.-B. Is there any relationship between dietary patterns and depression and anxiety in Chinese adolescents? *Public Health Nutr.* **2012**, *15*, 673–682. [CrossRef]

93. Marangoni, F.; Martini, D.; Scaglioni, S.; Sculati, M.; Donini, L.M.; Leonardi, F.; Agostoni, C.; Castelnuovo, G.; Ferrara, N.; Ghiselli, A. Snacking in nutrition and health. *Int. J. Food Sci. Nutr.* **2019**. [CrossRef] [PubMed]

94. Smith, R.G.; Betancourt, L.; Sun, Y. Molecular endocrinology and physiology of the aging central nervous system. *Endocr. Rev.* **2005**, *26*, 203–250. [CrossRef] [PubMed]

95. Takeda, E.; Terao, J.; Nakaya, Y.; Miyamoto, K.-I.; Baba, Y.; Chuman, H.; Kaji, R.; Ohmori, T.; Rokutan, K. Stress control and human nutrition. *Int. J. Med. Investig.* **2004**, *51*, 139–145. [CrossRef]

96. Spencer, J.P. The impact of fruit flavonoids on memory and cognition. *Br. J. Nutr.* **2010**, *104*, S40–S47. [CrossRef] [PubMed]

97. Bensalem, J.; Dudonné, S.; Gaudout, D.; Servant, L.; Calon, F.; Desjardins, Y.; Layé, S.; Lafenetre, P.; Pallet, V. Polyphenol-rich extract from grape and blueberry attenuates cognitive decline and improves neuronal function in aged mice. *J. Nutr. Sci.* **2018**, *7*. [CrossRef]

98. Bensalem, J.; Dudonné, S.; Etchamendy, N.; Pellay, H.; Amadieu, C.; Gaudout, D.; Dubreuil, S.; Paradis, M.-E.; Pomerleau, S.; Capuron, L. Polyphenols from grape and blueberry improve episodic memory in healthy elderly with lower level of memory performance: A bicentric double-blind, randomized, placebo-controlled clinical study. *J. Gerontol.: Ser. A* **2018**, *74*, 996–1007. [CrossRef]

99. Parmar, H.S.; Dixit, Y.; Kar, A. Fruit and vegetable peels: Paving the way towards the development of new generation therapeutics. *Drug Discov. Ther.* **2010**, *4*, 314–325.

100. Joseph, J.A.; Shukitt-Hale, B.; Denisova, N.A.; Bielinski, D.; Martin, A.; McEwen, J.J.; Bickford, P.C. Reversals of age-related declines in neuronal signal transduction, cognitive, and motor behavioral deficits with blueberry, spinach, or strawberry dietary supplementation. *J. Neurosci.* **1999**, *19*, 8114–8121. [CrossRef]

101. Miller, M.G.; Hamilton, D.A.; Joseph, J.A.; Shukitt-Hale, B. Dietary blueberry improves cognition among older adults in a randomized, double-blind, placebo-controlled trial. *Eur. J. Nutr.* **2018**, *57*, 1169–1180. [CrossRef]

102. Watson, A.; Okello, E.; Brooker, H.; Lester, S.; McDougall, G.; Wesnes, K. The impact of blackcurrant juice on attention, mood and brain wave spectral activity in young healthy volunteers. *Nutr. Neurosci.* **2018**, *22*, 596–606. [CrossRef]

103. Riaz, A.; Khan, R.A.; Algahtani, H.A. Memory boosting effect of Citrus limon, Pomegranate and their combinations. *Pak. J. Pharm. Sci.* **2014**, *27*, 1837–1840. [PubMed]

104. Rendeiro, C.; Guerreiro, J.D.; Williams, C.M.; Spencer, J.P. Flavonoids as modulators of memory and learning: Molecular interactions resulting in behavioural effects. *Proc. Nutr. Soc.* **2012**, *71*, 246–262. [CrossRef] [PubMed]

105. Esteban, J.A.; Shi, S.-H.; Wilson, C.; Nuriya, M.; Huganir, R.L.; Malinow, R. PKA phosphorylation of AMPA receptor subunits controls synaptic trafficking underlying plasticity. *Nat. Neurosci.* **2003**, *6*, 136. [CrossRef] [PubMed]

106. Kelleher, R.J., III; Govindarajan, A.; Jung, H.-Y.; Kang, H.; Tonegawa, S. Translational control by MAPK signaling in long-term synaptic plasticity and memory. *Cell* **2004**, *116*, 467–479. [CrossRef]

107. Kim, D.H.; Jeon, S.J.; Son, K.H.; Jung, J.W.; Lee, S.; Yoon, B.H.; Choi, J.W.; Cheong, J.H.; Ko, K.H.; Ryu, J.H. Effect of the flavonoid, oroxylin A, on transient cerebral hypoperfusion-induced memory impairment in mice. *Pharmacol. Biochem. Behav.* **2006**, *85*, 658–668. [CrossRef] [PubMed]

108. Jaeger, B.N.; Parylak, S.L.; Gage, F.H. Mechanisms of dietary flavonoid action in neuronal function and neuroinflammation. *Mol. Aspects Med.* **2018**, *61*, 50–62. [CrossRef]

109. Duan, Y.; Wang, Z.; Zhang, H.; He, Y.; Fan, R.; Cheng, Y.; Sun, G.; Sun, X. Extremely low frequency electromagnetic field exposure causes cognitive impairment associated with alteration of the glutamate level, MAPK pathway activation and decreased CREB phosphorylation in mice hippocampus: Reversal by procyanidins extracted from the lotus seedpod. *Food Funct.* **2014**, *5*, 2289–2297.

110. Johnston, G.A. Flavonoid nutraceuticals and ionotropic receptors for the inhibitory neurotransmitter GABA. *Neurochem. Int.* **2015**, *89*, 120–125. [CrossRef]

111. Clarkson, A.N.; Boothman-Burrell, L.; Dósa, Z.; Nagaraja, R.Y.; Jin, L.; Parker, K.; van Nieuwenhuijzen, P.S.; Neumann, S.; Gowing, E.K.; Gavande, N. The flavonoid, 2′-methoxy-6-methylflavone, affords neuroprotection following focal cerebral ischaemia. *J. Cereb. Blood Flow Metab.* **2018**. [CrossRef]

112. Gomez-Pinilla, F.; Nguyen, T.T. Natural mood foods: The actions of polyphenols against psychiatric and cognitive disorders. *Nutr. Neurosci.* **2012**, *15*, 127–133. [CrossRef]

113. Ma, Y.; Wang, H.; Wu, H.; Wei, C.; Lee, E. Brain-derived neurotrophic factor antisense oligonucleotide impairs memory retention and inhibits long-term potentiation in rats. *Neuroscience* **1997**, *82*, 957–967. [CrossRef]

114. Rahvar, M.; Owji, A.; Mashayekhi, F. Effect of quercetin on the brain-derived neurotrophic factor gene expression in the rat brain. *Bratisl. Lek. Listy* **2018**, *119*, 28–31. [CrossRef] [PubMed]

115. Gite, S.; Ross, R.P.; Kirke, D.; Guihéneuf, F.; Aussant, J.; Stengel, D.B.; Dinan, T.G.; Cryan, J.F.; Stanton, C. Nutraceuticals to promote neuronal plasticity in response to corticosterone-induced stress in human neuroblastoma cells. *Nutr. Neurosci.* **2019**, *22*, 551–568. [CrossRef] [PubMed]

116. Zuccato, C.; Cattaneo, E. Brain-derived neurotrophic factor in neurodegenerative diseases. *Nat. Rev. Neurol.* **2009**, *5*, 311. [CrossRef]

117. Alzoubi, K.H.; Alibbini, S.; Khabour, O.F.; El-Elimat, T.; Al-zubi, M.; Alali, F.Q. Carob (*Ceratonia siliqua* L.) Prevents Short-Term Memory Deficit Induced by Chronic Stress in Rats. *J. Mol. Neurosci.* **2018**, *66*, 314–321. [CrossRef]

118. Ko, Y.-H.; Kwon, S.-H.; Lee, S.-Y.; Jang, C.-G. Liquiritigenin ameliorates memory and cognitive impairment through cholinergic and BDNF pathways in the mouse hippocampus. *Arch. Pharmacal Res.* **2017**, *40*, 1209–1217. [CrossRef]

119. Swaab, D.F.; Bao, A.-M.; Lucassen, P.J. The stress system in the human brain in depression and neurodegeneration. *Ageing Res. Rev.* **2005**, *4*, 141–194. [CrossRef]

120. Francis, S.; Head, K.; Morris, P.; Macdonald, I. The effect of flavanol-rich cocoa on the fMRI response to a cognitive task in healthy young people. *J. Cardiovasc. Pharmacol.* **2006**, *47*, S215–S220. [CrossRef]

121. Bell, L.; Lamport, D.J.; Butler, L.T.; Williams, C.M. A review of the cognitive effects observed in humans following acute supplementation with flavonoids, and their associated mechanisms of action. *Nutrients* **2015**, *7*, 10290–10306. [CrossRef]

122. Rendeiro, C.; Rhodes, J.S.; Spencer, J.P. The mechanisms of action of flavonoids in the brain: Direct versus indirect effects. *Neurochem. Int.* **2015**, *89*, 126–139. [CrossRef]

123. Williams, C.M.; El Mohsen, M.A.; Vauzour, D.; Rendeiro, C.; Butler, L.T.; Ellis, J.A.; Whiteman, M.; Spencer, J.P. Blueberry-induced changes in spatial working memory correlate with changes in hippocampal CREB phosphorylation and brain-derived neurotrophic factor (BDNF) levels. *Free Radic. Biol. Med.* **2008**, *45*, 295–305. [CrossRef] [PubMed]

124. Van Praag, H.; Lucero, M.J.; Yeo, G.W.; Stecker, K.; Heivand, N.; Zhao, C.; Yip, E.; Afanador, M.; Schroeter, H.; Hammerstone, J. Plant-derived flavanol (−)Epicatechin enhances angiogenesis and retention of spatial memory in mice. *J. Neurosci.* **2007**, *27*, 5869–5878. [CrossRef] [PubMed]

125. Drouin, A.; Bolduc, V.; Thorin-Trescases, N.; Bélanger, É.; Fernandes, P.; Baraghis, E.; Lesage, F.; Gillis, M.-A.; Villeneuve, L.; Hamel, E. Catechin treatment improves cerebrovascular flow-mediated dilation and learning abilities in atherosclerotic mice. *Am. J. Physiol. Heart Circ. Physiol.* **2010**, *300*, H1032–H1043. [CrossRef] [PubMed]

126. Bisson, J.-F.; Nejdi, A.; Rozan, P.; Hidalgo, S.; Lalonde, R.; Messaoudi, M. Effects of long-term administration of a cocoa polyphenolic extract (Acticoa powder) on cognitive performances in aged rats. *Br. J. Nutr.* **2008**, *100*, 94–101. [CrossRef]

127. Joris, P.J.; Mensink, R.P.; Adam, T.C.; Liu, T.T. Cerebral Blood Flow Measurements in Adults: A Review on the Effects of Dietary Factors and Exercise. *Nutrients* **2018**, *10*, 530. [CrossRef] [PubMed]

128. Spencer, J.P.; Vauzour, D.; Rendeiro, C. Flavonoids and cognition: The molecular mechanisms underlying their behavioural effects. *Arch. Biochem. Biophys.* **2009**, *492*, 1–9. [CrossRef] [PubMed]

129. Xia, S.-F.; Xie, Z.-X.; Qiao, Y.; Li, L.-R.; Cheng, X.-R.; Tang, X.; Shi, Y.-H.; Le, G.-W. Differential effects of quercetin on hippocampus-dependent learning and memory in mice fed with different diets related with oxidative stress. *Physiol. Behav.* **2015**, *138*, 325–331. [CrossRef]

130. Qi, G.; Mi, Y.; Wang, Y.; Li, R.; Huang, S.; Li, X.; Liu, X. Neuroprotective action of tea polyphenols on oxidative stress-induced apoptosis through the activation of the TrkB/CREB/BDNF pathway and Keap1/Nrf2 signaling pathway in SH-SY5Y cells and mice brain. *Food Funct.* **2017**, *8*, 4421–4432. [CrossRef]

131. Ding, M.-L.; Ma, H.; Man, Y.-G.; Lv, H.-Y. Protective effects of a green tea polyphenol, epigallocatechin-3-gallate, against sevoflurane-induced neuronal apoptosis involve regulation of CREB/BDNF/TrkB and PI3K/Akt/mTOR signalling pathways in neonatal mice. *Can. J. Biochem. Physiol.* **2017**, *95*, 1396–1405. [CrossRef]

132. Zhao, X.; Li, R.; Jin, H.; Jin, H.; Wang, Y.; Zhang, W.; Wang, H.; Chen, W. Epigallocatechin-3-gallate confers protection against corticosterone-induced neuron injuries via restoring extracellular signal-regulated kinase 1/2 and phosphatidylinositol-3 kinase/protein kinase B signaling pathways. *PLoS ONE* **2018**, *13*, e0192083. [CrossRef]

133. Mi, Y.; Qi, G.; Fan, R.; Qiao, Q.; Sun, Y.; Gao, Y.; Liu, X. EGCG ameliorates high-fat–and high-fructose–induced cognitive defects by regulating the IRS/AKT and ERK/CREB/BDNF signaling pathways in the CNS. *FASEB J.* **2017**, *31*, 4998–5011. [CrossRef] [PubMed]

134. Wightman, E.L.; Haskell, C.F.; Forster, J.S.; Veasey, R.C.; Kennedy, D.O. Epigallocatechin gallate, cerebral blood flow parameters, cognitive performance and mood in healthy humans: A double-blind, placebo-controlled, crossover investigation. *Hum. Psychopharmacol.* **2012**, *27*, 177–186. [CrossRef] [PubMed]

135. Lamport, D.J.; Pal, D.; Macready, A.L.; Barbosa-Boucas, S.; Fletcher, J.M.; Williams, C.M.; Spencer, J.P.; Butler, L.T. The effects of flavanone-rich citrus juice on cognitive function and cerebral blood flow: An acute, randomised, placebo-controlled cross-over trial in healthy, young adults. *Br. J. Nutr.* **2016**, *116*, 2160–2168. [CrossRef] [PubMed]

136. Nehlig, A. The neuroprotective effects of cocoa flavanol and its influence on cognitive performance. *Br. J. Clin. Pharmacol.* **2013**, *75*, 716–727. [CrossRef] [PubMed]

137. Fisher, N.D.; Hughes, M.; Gerhard-Herman, M.; Hollenberg, N.K. Flavanol-rich cocoa induces nitric-oxide-dependent vasodilation in healthy humans. *J. Hypertens.* **2003**, *21*, 2281–2286. [CrossRef]

138. Karim, M.; McCormick, K.; Kappagoda, C.T. Effects of cocoa extracts on endothelium-dependent relaxation. *J. Nutr.* **2000**, *130*, 2105S–2108S. [CrossRef]

139. Hollenberg, N.K.; Fisher, N.D.; McCullough, M.L. Flavanols, the Kuna, cocoa consumption, and nitric oxide. *J. Am. Soc. Hypertens.* **2009**, *3*, 105–112. [CrossRef]

140. Decroix, L.; Tonoli, C.; Soares, D.D.; Tagougui, S.; Heyman, E.; Meeusen, R. Acute cocoa flavanol improves cerebral oxygenation without enhancing executive function at rest or after exercise. *Appl. Physiol. Nutr. Metab.* **2016**, *41*, 1225–1232. [CrossRef]

141. Kennedy, D.O.; Wightman, E.L.; Reay, J.L.; Lietz, G.; Okello, E.J.; Wilde, A.; Haskell, C.F. Effects of resveratrol on cerebral blood flow variables and cognitive performance in humans: A double-blind, placebo-controlled, crossover investigation. *Am. J. Clin. Nutr.* **2010**, *91*, 1590–1597. [CrossRef]

142. Goyarzu, P.; Malin, D.H.; Lau, F.C.; Taglialatela, G.; Moon, W.D.; Jennings, R.; Moy, E.; Moy, D.; Lippold, S.; Shukitt-Hale, B. Blueberry supplemented diet: Effects on object recognition memory and nuclear factor-kappa B levels in aged rats. *Nutr. Neurosci.* **2004**, *7*, 75–83. [CrossRef]

143. Chen, T.; Yang, Y.-J.; Li, Y.-K.; Liu, J.; Wu, P.-F.; Wang, F.; Chen, J.-G.; Long, L.-H. Chronic administration tetrahydroxystilbene glucoside promotes hippocampal memory and synaptic plasticity and activates ERKs, CaMKII and SIRT1/miR-134 in vivo. *J. Ethnopharmacol.* **2016**, *190*, 74–82. [CrossRef] [PubMed]

144. Lu, Y.; Sareddy, G.R.; Wang, J.; Wang, R.; Li, Y.; Dong, Y.; Zhang, Q.; Liu, J.; O'Connor, J.C.; Xu, J. Neuron-Derived Estrogen Regulates Synaptic Plasticity and Memory. *J. Neurosci.* **2019**, *39*, 2792–2809. [CrossRef] [PubMed]

145. Mancini, E.; Beglinger, C.; Drewe, J.; Zanchi, D.; Lang, U.E.; Borgwardt, S. Green tea effects on cognition, mood and human brain function: A systematic review. *Phytomedicine* **2017**, *34*, 26–37. [CrossRef] [PubMed]

146. Pervin, M.; Unno, K.; Takagaki, A.; Isemura, M.; Nakamura, Y. Function of Green Tea Catechins in the Brain: Epigallocatechin Gallate and its Metabolites. *Int. J. Mol. Sci.* **2019**, *20*, 3630. [CrossRef]

147. Kakutani, S.; Watanabe, H.; Murayama, N. Green tea intake and risks for dementia, Alzheimer's disease, mild cognitive impairment, and cognitive impairment: A systematic review. *Nutrients* **2019**, *11*, 1165. [CrossRef]

148. Babaei, F.; Mirzababaei, M.; Nassiri-Asl, M. Quercetin in Food: Possible Mechanisms of Its Effect on Memory. *J. Food Sci.* **2018**, *83*, 2280–2287. [CrossRef]

149. Karimipour, M.; Rahbarghazi, R.; Tayefi, H.; Shimia, M.; Ghanadian, M.; Mahmoudi, J.; Bagheri, H.S. Quercetin promotes learning and memory performance concomitantly with neural stem/progenitor cell proliferation and neurogenesis in the adult rat dentate gyrus. *Int. J. Dev. Neurosci.* **2019**, *74*, 18–26. [CrossRef]

150. Ishola, I.O.; Osele, M.O.; Chijioke, M.C.; Adeyemi, O.O. Isorhamnetin enhanced cortico-hippocampal learning and memory capability in mice with scopolamine-induced amnesia: Role of antioxidant defense, cholinergic and BDNF signaling. *Brain Res.* **2019**, *1712*, 188–196. [CrossRef]

151. Brickman, A.M.; Khan, U.A.; Provenzano, F.A.; Yeung, L.-K.; Suzuki, W.; Schroeter, H.; Wall, M.; Sloan, R.P.; Small, S.A. Enhancing dentate gyrus function with dietary flavanols improves cognition in older adults. *Nat. Neurosci.* **2014**, *17*, 1798. [CrossRef]

152. Wightman, E.L.; Haskell-Ramsay, C.F.; Reay, J.L.; Williamson, G.; Dew, T.; Zhang, W.; Kennedy, D.O. The effects of chronic trans-resveratrol supplementation on aspects of cognitive function, mood, sleep, health and cerebral blood flow in healthy, young humans. *Br. J. Nutr.* **2015**, *114*, 1427–1437. [CrossRef]

153. Scholey, A.B.; French, S.J.; Morris, P.J.; Kennedy, D.O.; Milne, A.L.; Haskell, C.F. Consumption of cocoa flavanols results in acute improvements in mood and cognitive performance during sustained mental effort. *J. Psychopharmacol.* **2010**, *24*, 1505–1514. [CrossRef] [PubMed]

154. Zainuddin, M.S.A.; Thuret, S. Nutrition, adult hippocampal neurogenesis and mental health. *Br. Med. Bull.* **2012**, *103*, 89. [CrossRef] [PubMed]

155. Ward, E. Addressing nutritional gaps with multivitamin and mineral supplements. *Nutr. J.* **2014**, *13*, 72. [CrossRef] [PubMed]

156. Macready, A.L.; Kennedy, O.B.; Ellis, J.A.; Williams, C.M.; Spencer, J.P.; Butler, L.T. Flavonoids and cognitive function: A review of human randomized controlled trial studies and recommendations for future studies. *Genes Nutr.* **2009**, *4*, 227. [CrossRef] [PubMed]

157. Almeida, M.; Pontes, F.; Barcelos, M.; Rosa, J.; Cruz, R. Antioxidant effect of flavonoids present in Euterpe oleracea Martius and neurodegenerative diseases: A literature review. *Cent. Nerv. Syst. Agents Med. Chem.* **2019**, *19*, 75–99.

158. Auti, S.T.; Kulkarni, Y.A. A systematic review on the role of natural products in modulating the pathways in Alzheimer's disease. *Int. J. Vitam. Nutr. Res.* **2017**, *87*, 99–116. [CrossRef]

159. Ayaz, M.; Sadiq, A.; Junaid, M.; Ullah, F.; Ovais, M.; Ullah, I.; Ahmed, J.; Shahid, M. Flavonoids as prospective neuroprotectants and their therapeutic propensity in aging associated neurological disorders. *Front. Aging Neurosci.* **2019**, *11*. [CrossRef]

160. Yates, A.A.; Erdman Jr, J.W.; Shao, A.; Dolan, L.C.; Griffiths, J.C. Bioactive nutrients-time for tolerable upper intake levels to address safety. *Regul. Toxicol. Pharmacol.* **2017**, *84*, 94–101. [CrossRef]

161. Santos-Buelga, C.; González-Paramás, A.M.; Oludemi, T.; Ayuda-Durán, B.; González-Manzano, S. Plant phenolics as functional food ingredients. *Adv. Food Nutr. Res.* **2019**, *90*, 183–257.

162. Salehi, B.; Iriti, M.; Vitalini, S.; Antolak, H.; Pawlikowska, E.; Kręgiel, D.; Sharifi-Rad, J.; Oyeleye, S.I.; Ademiluyi, A.O.; Czopek, K. Euphorbia-Derived Natural Products with Potential for Use in Health Maintenance. *Biomolecules* **2019**, *9*, 337. [CrossRef]

163. Piccolella, S.; Crescente, G.; Candela, L.; Pacifico, S. Nutraceutical Polyphenols: New analytical challenges and opportunities. *J. Pharm. Biomed. Anal.* **2019**, *175*. [CrossRef] [PubMed]

164. Deardorff, W.J.; Feen, E.; Grossberg, G.T. The use of cholinesterase inhibitors across all stages of Alzheimer's disease. *Drugs Aging* **2015**, *32*, 537–547. [CrossRef] [PubMed]

165. Miroddi, M.; Navarra, M.; Quattropani, M.C.; Calapai, F.; Gangemi, S.; Calapai, G. Systematic Review of Clinical Trials Assessing Pharmacological Properties of S alvia Species on Memory, Cognitive Impairment and A lzheimer's Disease. *CNS Neurosci. Ther.* **2014**, *20*, 485–495. [CrossRef] [PubMed]

166. Flanagan, E.; Müller, M.; Hornberger, M.; Vauzour, D. Impact of Flavonoids on Cellular and Molecular Mechanisms Underlying Age-Related Cognitive Decline and Neurodegeneration. *Curr. Nutr. Rep.* **2018**, *7*, 49–57. [CrossRef] [PubMed]
167. Lamport, D.J.; Saunders, C.; Butler, L.T.; Spencer, J.P. Fruits, vegetables, 100% juices, and cognitive function. *Nutr. Rev.* **2014**, *72*, 774–789. [CrossRef]

MDPI

St. Alban-Anlage 66

4052 Basel

Switzerland

Tel. +41 61 683 77 34

Fax +41 61 302 89 18

www.mdpi.com

Foods Editorial Office

E-mail: foods@mdpi.com

www.mdpi.com/journal/foods